Materiomics

This complete, yet concise, guide introduces you to the rapidly developing field of high-throughput screening of biomaterials: materiomics. Bringing together the key concepts and methodologies used to determine biomaterial properties, it will allow you to understand the adaptation and application of materiomics in areas such as rapid prototyping, lithography and combinatorial chemistry. Each chapter is written by internationally renowned experts, and includes tutorial paragraphs on topics such as biomaterial-banking, imaging, assay development, translational aspects and informatics. Case studies of state-of-the-art experiments provide illustrative examples, while lists of key publications allow you to read up easily on the most relevant background material. Whether you are a professional scientist in industry, a student or a researcher, this book is not to be missed if you are interested in the latest developments in biomaterials research.

Jan de Boer is a Professor of Applied Cell Biology at the University of Twente, the Netherlands, at the MIRA Institute for Biomedical Technology and Technical Medicine, where his team performs innovative research on molecular and cellular engineering of bone tissue. He is chair of the Netherlands Society of Biomaterials and Tissue Engineering, and co-founder of the biotech company Materiomics B.V.

Clemens A. van Blitterswijk is Professor of Tissue Regeneration at the University of Twente, and chair of the Department of Tissue Regeneration at MIRA. Clemens has published over 300 papers and is co-inventor on over 100 patents. He is one of only two individuals to have received both the Jean Leray Award and the George Winter Award, the two prestigious prizes of the European Society for Biomaterials. He was recently appointed as a fellow of the Royal Netherlands Academy of Sciences and is a fellow of the Netherlands Academy of Technology and Innovation.

Materiomics

High-Throughput Screening of Biomaterial Properties

JAN DE BOER
University of Twente

CLEMENS A. VAN BLITTERSWIJK
University of Twente

CAMBRIDGE UNIVERSITY PRESS
Cambridge, New York, Melbourne, Madrid, Cape Town,
Singapore, São Paulo, Delhi, Mexico City

Cambridge University Press
The Edinburgh Building, Cambridge CB2 8RU, UK

Published in the United States of America by Cambridge University Press, New York

www.cambridge.org
Information on this title: www.cambridge.org/9781107016774

First published 2013

Printed and bound in the United Kingdom by the MPG Books Group

A catalogue record for this publication is available from the British Library

Library of Congress Cataloguing in Publication data
Materiomics : high-throughput screening of biomaterial properties / [edited by] Jan de Boer,
Clemens van Blitterswijk.
   p. ; cm.
Includes bibliographical references.
ISBN 978-1-107-01677-4 (hardback)
I. Boer, Jan de, 1970–   II. Blitterswijk, Clemens A. van.
[DNLM: 1. Biocompatible Materials – chemistry.   2. Biocompatible Materials – analysis.   3. Materials
Testing.   4. Regenerative Medicine – methods.   5. Tissue Engineering – methods.   QT 37]
610.28′4–dc23

                                                                                            2012038878

ISBN 978-1-107-01677-4 Hardback

Contents

Contributors

Beckham, Haskell
Georgia Institute of Technology, School of Materials Science and
Engineering, Atlanta, USA

Beijersbergen, Roderick
The Netherlands Cancer Institute, Division Molecular Carcinogenesis,
Amsterdam, The Netherlands

van Blitterswijk, Clemens A.
Department of Tissue Regeneration, MIRA Institute for Biomedical
Technology and Technical Medicine, University of Twente,
Enschede, The Netherlands

de Boer, Jan
Department of Tissue Regeneration, MIRA Institute for Biomedical
Technology and Technical Medicine, University of Twente,
Enschede, The Netherlands

Bosman, Mirjam
Department of Biomaterials Science and Technology, University of Twente,
Enschede, The Netherlands

Bradley, Mark
School of Chemistry, University of Edinburgh, Edinburgh, UK

Buehler, Markus
Department of Civil and Environmental Engineering, MIT, Cambridge, USA

Cranford, Steven
Department of Civil and Environmental Engineering, MIT, Cambridge, USA

Dankers, Patricia
Institute for Complex Molecular Systems and Laboratory of Chemical Biology,
Eindhoven University of Technology, Eindhoven, The Netherlands

Danoux, Charlène
Department of Tissue Regeneration, MIRA Institute for Biomedical
Technology and Technical Medicine, University of Twente,
Enschede, The Netherlands

Dechering, Koen
TropIQ Health Sciences, Nijmegen, The Netherlands

Félix Lanao, Rosa
Department of Biomaterials, Radboud University Nijmegen Medical Centre,
Nijmegen, The Netherlands

Fernandes, Hugo
Department of Tissue Regeneration, MIRA Institute for Biomedical
Technology and Technical Medicine, University of Twente,
Enschede, The Netherlands

Ferreira, Lino
CNC – Centre for Neuroscience and Cell Biology, University of Coimbra,
Coimbra, Portugal; Biocant – Centre of Innovation and Biotechnology,
Cantanhede, Portugal

Gauvin, Robert
Harvard–MIT Division of Health Sciences and Techology, Boston, MA, USA;
Quebec Centre for Functional Materials, Université Laval, Québec City, QC, Canada

Geris, Liesbet
Biomechanics Research Unit, University of Liège, Belgium Prometheus,
KU Leuven R&D division for skeletal tissue engineering, KU Leuven, Belgium

Gobaa, Samy
Laboratory of Stem Cell Bioengineering and Institute of Bioengineering,
Ecole Polytechnique Fédérale de Lausanne, Lausanne, Switzerland

Gomez, Ismael
Georgia Institute of Technology, School of Chemical and Biomolecular
Engineering, Atlanta, USA

Grijpma, Dirk
Department of Biomaterials Science and Technology, MIRA Institute for
Biomedical Technology and Technical Medicine, University of Twente,
Enschede, The Netherlands; Department of Biomedical Engineering, University
Medical Centre Groningen, Groningen, The Netherlands

Groen, Nathalie
Department of Tissue Regeneration, MIRA Institute for Biomedical
Technology and Technical Medicine, University of Twente, Enschede,
The Netherlands

Guillemot, Fabien
INSERM, University of Bordeaux, Bordeaux, France

Habibovic, Pamela
Department of Tissue Regeneration, MIRA Institute for Biomedical Technology
and Technical Medicine, University of Twente, Enschede, The Netherlands

Han, Sangil
Georgia Institute of Technology, School of Chemical and Biomolecular
Engineering, Atlanta, USA

Higuera, Gustavo
Department of Tissue Regeneration, MIRA Institute for Biomedical Technology
and Technical Medicine, University of Twente, Enschede, The Netherlands

Hulshof, Frits
Department of Tissue Regeneration, MIRA Institute for Biomedical Technology
and Technical Medicine, University of Twente, Enschede, The Netherlands

Hulsman, Marc
Delft Bioinformatics Lab, Delft University of Technology, Delft, The Netherlands

Hunt, John
Clinical Engineering (UK CTE), UKBioTEC, Institute of Ageing and Chronic
Disease, University of Liverpool, Liverpool, UK

Khademhosseini, Ali
Harvard–MIT Division of Health Sciences and Technology, Massachusetts
Institute of Technology, Cambridge, USA

Leeuwenburgh, Sander
Department of Biomaterials, Radboud University Nijmegen Medical Center,
Nijmegen, The Netherlands

Leisen, Johannes
Georgia Institute of Technology, School of Chemistry and Biochemistry, Atlanta, USA

Liu, Er
Department of Biomedical Engineering, La Jolla Bioengineering Institute, San Diego, USA

Lutolf, Matthias
Laboratory of Stem Cell Bioengineering and Institute of Bioengineering, Ecole Polytechnique Fédérale de Lausanne, Lausanne, Switzerland

Mahmood, Tahir
Life Sciences Practice – Commercial Healthcare Group, Booz Allen Hamilton, San Francisco, USA

Maniura-Weber, Katharina
Empa – Swiss Federal Laboratories for Materials Science and Technology, Laboratories for Materials–Biology Interactions, St Gallen, Switzerland

Meredith, Carson
Georgia Institute of Technology, School of Chemical and Biomolecular Engineering, Atlanta, USA

Moghe, Prabhas
Department of Biomedical Engineering, Rutgers University, Piscataway, New Jersey, USA; Department of Chemical & Biochemical Engineering, Rutgers University, Piscataway, New Jersey, USA

Mollet, Björne
Institute for Complex Molecular Systems and Laboratory of Chemical Biology, Eindhoven University of Technology, Eindhoven, The Netherlands

Moroni, Lorenzo
Department of Tissue Regeneration, MIRA Institute for Biomedical Technology and Technical Medicine, University of Twente, Enschede, The Netherlands

Negro, Andrea
Laboratory of Stem Cell Bioengineering and Institute of Bioengineering, Ecole Polytechnique Fédérale de Lausanne, Lausanne, Switzerland

Oreffo, Richard
Bone and Joint Research Group, Centre for Human Development, Stem Cells and Regeneration, Human Development and Health, Institute of Developmental Sciences University of Southampton, Southampton, UK

Pape, Bram
Institute for Complex Molecular Systems and Laboratory of Chemical Biology, Eindhoven University of Technology, Eindhoven, The Netherlands

Pernagallo, Salvatore
School of Chemistry, University of Edinburgh, Edinburgh, UK

Reinders, Marcel
Delft Bioinformatics Lab, Delft University of Technology, Delft, The Netherlands

Smith, James
Bone & Joint Research Group, Centre for Human Development, Stem Cells and Regeneration, Human Development and Health, Institute of Developmental Sciences University of Southampton, Southampton, UK

Tare, Rahul
Bone & Joint Research Group, Centre for Human Development, Stem Cells and Regeneration, Human Development and Health, Institute of Developmental Sciences University of Southampton, Southampton, UK

Truckenmüller, Roman
Department of Tissue Regeneration, MIRA Institute for Biomedical Technology and Technical Medicine, University of Twente, Enschede, The Netherlands

Unadkat, Hemant
Mechanobiology Institute, National University of Singapore, Singapore

Zant, Erwin
Department of Biomaterials Science and Technology, University of Twente, Enschede, The Netherlands

Zhang, Rong
School of Chemistry, University of Edinburgh, Edinburgh, UK

Preface

It's that sense of unease when you step out of the airport terminal building and onto the streets of Kathmandu. Or the moment when you open the door to your new office to see unfamiliar faces waiting for you. Step out of your comfort zone and discover how exciting, thrilling and liberating it can be: a new world is waiting for you. This book is about stepping out of the comfort zone of your own scientific discipline and about exposing yourself to something new. Embrace all the scientific disciplines that build modern-day biomaterials research, in the cultural hotpot of materiomics. Don't let the jargon and three-letter abbreviations of cell biology hold you back, nor the abracadabra of statistical models, nor the Latin terms for body parts and diseases. Learn a new language and a whole new culture is waiting for you.

The compilation of this book was initiated after an exciting conference termed 'High throughput screening of biomaterials: shaping a new research area', held beside the Amsterdam canals in April 2011. The meeting was attended by 50 selected scientists from all over the globe and across all the disciplines of biomaterials research, and the format of the conference took away that sense of unease. Chemists talked to clinicians, biologists listened to information scientists, engineers brainstormed with policy makers. We decided to bring this open and inviting atmosphere to the public through this book. Therefore, each chapter contains a tutorial on the topic for non-experts, gives an overview of the current status of that field and discusses how this technology will further shape the future of materiomics. The result of this exciting journey is presented here and was made possible only with the help of all the authors and those who contributed to the organization of the conference (Anouk Mentink), the editing of the book (Ruben Burer) or the chapters (Kristen Johnson). We hope that this book will be a scientific passport which lets you travel across the border of your discipline and helps you to learn to appreciate that of others. You won't be disappointed. Enjoy your journey!

Jan de Boer

'The adventurous spirit of this book, and indeed the field of materiomics, is excellently prefaced by Jan de Boer in this thorough compilation of concise chapters produced by an international cast of experts. It succeeds in its aim to be of use to both the student and the experienced practitioner in the multi-faceted emerging discipline of materiomics, containing both useful information and thought-provoking discussion and future perspectives. I would recommend it both to those interested in and to those already immersed in this rapidly evolving field.'

Morgan Alexander, The University of Nottingham

'By dissecting the contribution of various disciplines of diverse nature, ranging from chemistry to informatics or from advanced imaging to rapid prototyping, the book organically defines Materiomics as a field of its own. The reader is ultimately left with the awareness that the field of Materiomics will play a central role in future approaches to design complex material systems with predictable properties, for biomedical or industrial applications.'

Ivan Martin, University Hospital Basel

1 Introducing materiomics

Nathalie Groen, Steven W. Cranford, Jan de Boer,
Markus J. Buehler and Clemens A. Van Blitterswijk

1.1 Introduction to materiomics

The ability to regenerate and repair tissues and organs – using science and engineering to supplement biology – continuously intrigues and inspires those hoping that the frailty of our bodies can be ultimately avoided. From ancient times, a surprising range of unnatural materials have been used to (partially) substitute human tissues for medicinal purposes. For example, in the era of the Incas (*c.* 1500), moulded materials such as gold and silver were used for the 'surgical' repair of cranial defects. In addition, archaeological findings reveal a wide range of materials, such as bronze, wood and leather, being used to replace and repair parts of the human body. Continuous refinement led to the first evidence of materials successfully implanted *inside* the body, reportedly used to repair a bone defect in the seventeenth century (see Further Reading).

Even earlier than this, the relationships between anatomy (i.e. structure) and function of living systems had been explored by Leonardo da Vinci and Galileo Galilei, who were among the first few to apply fundamental science to biological systems. In the current age of technology, new materials for biomedical and clinical application have undergone a modern Renaissance, resulting in a surge in design and successful application (1–5). The concepts of tissue repair and substitution are constantly improving and becoming more accessible, as proven for example by the widespread occurrence (and popular approval) of total hip and knee replacements. But rather than replacement with synthetic analogues, can biological tissue(s) be directly engineered?

The first biomaterials arose to solve specific clinical problems, and it was only later that this became a field of research in itself. Polymers and ceramics (and other effective biomaterials) were not developed for implants *per se*, but rather were used because of their availability and proven (known) material properties. This need not be the case. The field of biomaterials has witnessed exciting and accelerating progression, partly owing to the emergence of physical-science-based approaches in the biological sciences. Consequently, developments have led to a number of blockbuster materials which currently play a substantial part in modern healthcare, with various clinical applications ranging from degradable intraocular lenses and sutures to coronary stents, heart valves and orthopaedic implants. But ultimately, where does this field lead?

Materiomics: High-Throughput Screening of Biomaterial Properties, ed. Jan de Boer and Clemens van Blitterswijk. Published by Cambridge University Press. © Cambridge University Press 2013.

1.2   The challenge of 'living' materials science

Hitherto, the field of biomaterials has largely been characterized by trial-and-error experimentation, practical intuition and low-throughput research (6). As a result, identification and development of successful biomaterial candidates has frequently been iterative, employing *ad hoc*, piece-wise or one-off approaches to design and characterize materials for a specific application (7). Currently lacking is a single set of 'design parameters' that can satisfy more than the most rudimentary system – there is neither a standard 'code' for biological systems nor a standard 'toolset' for analysis.

Despite continuous advances both in the understanding of the natural function of biological materials and systems and in the synthesis and regeneration of certain tissues (such as bone), a cohesive and systematic approach is still wanting. What is the primary impediment? Biological tissues, organs and materials exploit multiple structures and functions across scales – they are universally hierarchical (8, 9). Such multiscale *hierarchies* consequently make any single-scale analysis and prediction a hypothesis at best. While studies have successfully characterized components at specific scales (e.g. the molecular structure of DNA or the sequence of a multitude of proteins), superposition of the structure or the functional properties of individual components (defined differently according to scale) is insufficient to understand the complete system (10). In simpler terms, '1 + 1 ≠ 2'. We utterly fail in the 'design' and 'construction' of such material systems – we cannot accurately or reliably predict behaviour of the final product. Indeed, whether through a lack of critical system variables or understanding of system response, we are unable to model larger (living) multiprotein systems and networks, let alone the structural role that such materials play in a cellular tissue. This is the exact opposite of the definition of engineering, where it is necessary to prescribe the performance of system components with reliable and repeatable accuracy.

Conversely, understanding the interaction of materials with biological ('living') tissues across all scales – from atoms and molecules to tissues and eventually at the organism level – remains a crucial hurdle in tissue engineering and biomaterial development. The challenge is intrinsically double-sided, yet highly intertwined. The scientific complexity at both sides of the interface – the material on the one hand and the organism on the other – needs to be considered (Figure 1.1). The fundamental problem of combining living (biological) and non-living (synthetic) components can be encapsulated by the popular adage, 'The whole is greater than the sum of its parts' (commonly attributed to Aristotle, who probably was not referring to the interface of biology and materials). The complex interactions between materials and biological systems require a certain flair to analyse deterministic (or predictive) behaviours and material properties. Nature, through meticulous trial and error over centuries of optimization and refinement, has intricately combined material structure, properties and functionality (9). Structure and function are so intimately linked that one-to-one substitution of other potential materials is currently not possible – but need this be the case?

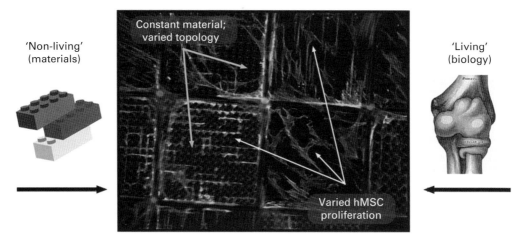

'Non-living'
(materials)

Constant material;
varied topology

'Living'
(biology)

Varied hMSC
proliferation

Figure 1.1   At the interface of materials and biology. The combination of living and non-living components –
namely biological (represented by a human knee joint) and synthetic materials (represented by
building blocks) – presents a complex challenge that can be summarized by the adage, 'The whole
is greater than the sum of its parts'. Here, the image shows the differential response of human
mesenchymal stem cells (hMSCs) to different underlying topographies on a so-called 'TopoChip'
(11). Image courtesy of Frits Hulshof.

1.3   Dealing with complexity

Clearly, the concepts of Nature cannot be omitted from the equation when developing
materials for biological applications. Evolutionary processes have resulted in intricate
biological systems, with robust and adaptable redundancies, as well as multifunctional
and multiscale components, which hamper compatible materials research – there are no
material 'standards' that all of biology must follow. This intrinsic complexity impedes
full understanding and limits developments in materials research for biological applications. Yet modern research has not sat idle, and has certainly led us to realize the *de facto*
complexity associated with biological systems. From a broad perspective, the causes of
this complexity can be grouped into common categories: multiscale; combinatorial and
temporal (see Figure 1.2).

    While the composition of biological materials is controlled by a relatively small set of
elements (carbon, hydrogen, oxygen, nitrogen and a few metal ions), this restriction is not
imposed on biomaterials research (yet the laws and principles of materials science and
chemistry remain applicable, allowing exploration beyond the confines of Nature).
Nature is highly successful in creating diversity from this limited set of 'building
blocks' – as was indisputably demonstrated by the discovery of the structure of DNA
by Watson and Crick in 1953, creating the illusion of a simple origin of life (relying on
only four nucleotides). As a result, the idea of growing any desired tissue from its basic
DNA code, along with emerging expertise in (biological) material processing, became
viable. One could foresee growing any desired tissue from the necessary DNA (along
with requisite raw materials), similar to the chemical vapour deposition of carbon

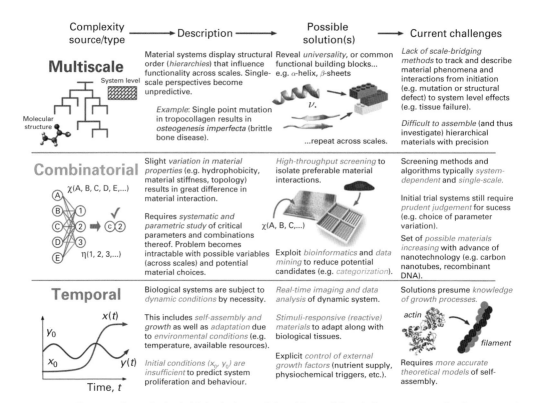

Figure 1.2   Sources of complexity in biological materials, with possible solutions *via* a materiomics approach.

nanotubes or the polymerization and spinning of nylon. It would merely be the assembly of the appropriate 'blocks', so to speak.

Yet Nature turned out to be more clever than that; evolution seamlessly intertwined structure and functionality. Despite protein materials being built, or 'transcribed', from a mere set of twenty amino acids, combinations of this limited set of building blocks produce a multitude of functionally distinct proteins (9). That being said, proteins acquire their functionality across multiple scales *via* a combination of peptide sequence and common structural motifs (such as α-helices and β-sheets) and a set of prevalent processes and mechanisms (e.g. synthesis, breakdown, self-assembly). The phenomenon of *universality* exists ubiquitously in biology. At higher scales, revealing the dimensions of biological complexity, proteins iteratively assemble into complexes: collagen fibrils, for example, which in turn form collagen fibres and eventually assemble together with additional inorganic materials, are the major constituents of bone tissue. The structural conformation of proteins might be highly conserved throughout different tissues, while concurrently (and contrastingly) being highly tissue-specific.

A key starting point in developing working models for such complex systems is the preservation of particular functionality despite uncertainty or minor variation in components and/or in the environment (10). We must neglect the physical idiosyncrasies of a system (such as specific peptide sequence), identify the fundamental building blocks (e.g. structure, key interacting groups) and delineate the function of each (signalling, catalytic, mechanical, etc.). In essence, biological systems originate from their associated genomic sequence – a distinct sequence of simple base pairs. While true, such a description is as crude as describing Beethoven as a simple collection of notes or the works of Shakespeare as a linear sequence of letters (12–14). The structural hierarchy and associated functionalities across scales add extra layers of complexity.

Another level of complexity arises from dynamic changes in biological material systems over time, owing to growth or adaption, for instance. To illustrate, the functional properties of proteins are also highly influenced by post-translational modifications (e.g. hydroxylations, phosphorylations, glycosylations) or enzymatic cross-linking. These modifications are crucial for interaction with other proteins and material components, and so determine the properties of tissues. At larger scales, cell adhesion, cytokinesis and cell migration illustrate the power of the cytoskeleton to self-organize locally into complex structures. This complexity impedes understanding of biological processes, as they are difficult to mimic or predict *ex vivo* or through synthetic approaches, posing a major challenge in structure prediction (and design) and the development of biocompatible materials. Simply put, biological materials grow (and/or evolve), while synthetic materials do not (they are characterized by static/constant material properties). It is apparent that not only are predictive models of assembly required, but also the possible development of self-adapting materials to mimic biological analogues.

Nature has creatively produced a broad range of functionally disparate materials (*diversity*) using a limited number of (*universal*) constituents, rather than inventing new building blocks. Such multiscale hierarchical systems simply cannot be analysed or predicted at a single scale. The so-called *universality–diversity paradigm* (15, 16) presents an alternative approach; it shifts the focus from individual component analysis towards the analysis of fundamental elements, hierarchical organization and functional mechanisms (sometimes referred to as *emergent* properties, a concept common in the scope of systems biology). Yet again, the whole is greater than the sum of its parts.

But how can we (a) determine what function is required and (b) reduce the number of potential material candidates for our need? Two main approaches can be considered in materials research to deal with this complexity and understand and engineer biological systems: firstly, via a bottom-up approach, identify fundamental building blocks and study their structure, interactions and properties at all relevant scales, from Ångstrom- to macro-level (from a single peptide to the collagen fibre); secondly, via a high-throughput approach, study the biological roles of a material system as a whole (combining the best of holistic and reductionist approaches; see Further Reading).

Within the first perspective, investing in the relation between universal structures and corresponding functions is similar to the field of proteomics (study of the function and structure of proteins) and interactomics (study of the web of interactions between biological molecules in a cell) (17–19), but extended beyond the confines of a cell and tissue to interactions and properties of materials. Observation and extraction of the general underlying principles (e.g. physical, chemical, optical, electronic, thermal, mechanical, etc.) of the structure–function relationship, using both experiments and theory, is required to make them available as concepts useful in materials science and engineering beyond biological occurrence, so that they should theoretically hold for similar synthetic material systems (20). But biological systems present inevitable complexity, introducing constraints in materials interactions analysis. Fields such as biomimetics attempt to exploit the structure and function (including the complexity) of such biological systems, applying principles of biology to synthetic systems for the design and engineering of material systems (20, 21).

Continuing this line of thought – applying biological 'tricks' to synthetic systems – we find that the problem quickly becomes intractable, as the sheer number of possible material–material interactions is unbounded. Moreover, unlike the biological limitation to available amino acids and ambient environmental conditions, in biomaterial research the complexity is further increased by the number of controls and variables produced by engineers (either by necessity or by choice).

The second approach mentioned above, the use of high-throughput combinatorial methods, may open up new possibilities. High-throughput-based methods allow simultaneous synthesis/processing and evaluating of a multitude of system variations (e.g. material, molecular) (22) to isolate desired behaviour/responses. Such methods have been commonly used in pharmacology for drug discovery (23), for the successful genetic screening of fruit flies and zebra fish (23) and for various applications in systems biology (24), to mention a few examples. Building on these past successes (also including proteomics or genomics (25, 26)), modern approaches have accelerated the discovery process and analytical methods, and have likewise extended insights and potential applications. Far from autonomous improvement, successful studies rely on technological advances in many fields, as every step involved in this approach requires high-throughput methods; from synthesis characterization (e.g. from a chemical or structural perspective) to analysis and characterization of the desired outcome (at cellular or tissue level) (27).

The screening process is relatively simple: when the desired performance is attained (based on a variety of metrics), a suitable material or system candidate can be defined, and subsequently iterated. The better candidates can then be investigated in more detail, to determine the relation between 'universal' material components and observed biological response, such as the relationship between surface chemistry or topology and a biological phenomenon of interest such as cell differentiation. The pathways and mechanisms thus unravelled may serve as a basis for further material refinement and development. An advantage is that no theoretical background of complex biological processes is required to screen for performance of material systems – only the results drive the screening process. Critical performance metrics and material properties may unexpectedly emerge upon characterization and analysis of successful outcomes, leading to new insights and target parameters. Such holistic screening of systems, together with reductionist characterization of the phenomenon, can be beneficial both for finding new systems and determining the mechanisms involved, providing a self-optimizing protocol for delineating material system characteristics and performance, beyond the scope of any one-off system investigation. High-throughput screening of a material property within a specific application can lead to unexpected findings, or even properties that could not have arisen naturally, which can in turn lead to optimized design of new materials.

In spite of the discussed intrinsic biological complexity, recent advances in (biological) material sciences have been considerable. Continuous refinement of techniques is providing new, more accurate means to measure, interpret, quantify and model the relationships between chemistry, structures, design and function. Progress in information technology, imaging, nanotechnology and related fields – coupled with developments in computing, modelling and simulation – has transformed investigative approaches of materials systems. The motivation has come from a vast assortment of disciplines: for example medicine (physiological properties of tissues for prosthetic devices, replacement materials and tissue engineering), biology (material aspects of adaptation, evolution, functionality, etc.) and materials science (thermal and electrical properties of nanosystems, functional performance of microscale devices, etc.). The potential reward and challenge of understanding biological materials elicits contributions from biologists, chemists and engineers alike. Further progress is hindered, however, by a 'divide and conquer' approach, and instead dictates a convergence of scientific disciplines under a common banner – and this is what is known as *materiomics*.

1.4. Emergence of materiomics

Traditionally, materials science, in its broadest sense, has been divided into distinct research areas based on classes of structures, length scales and varying functionalities (structural, thermal, electronic, etc.). Disparate disciplinary affiliations coexist, such as the specialities of ceramics and polymers, the fields of nano- and microtechnology, or the area of bioactive materials, for specific applications. In Nature, however, reciprocal refinement (i.e. 'evolution') has led to a balance between chemistry, materials, structure and required function. From this perspective, the disciplinary boundaries in material sciences should be razed, and the merger (or *convergence*) of different disciplines is inevitable. The rich history, experience and unique perspectives of distinct fields promote progress in this inherently interdisciplinary venture (Figure 1.3). Unsurprisingly, combining the widespread knowledge of materials scientists with the detailed understanding of biological systems and structures built over years by biologists holds great promise.

The emerging field of materiomics works from this philosophy of convergence and is characterized by an approach that considers all mechanisms of a material system across multiple scales. Materiomics – the transparent combination of 'material' with '-omics' – is most simply defined as the holistic study of materials systems. It approaches biological materials science (systems with or without synthetic components) through the integration of natural functions and biological processes ('living' interactions) with traditional materials science perspectives (physical properties, chemical components, hierarchical structures, mechanical behaviour, etc.). The suffix -omics, as in fields such as proteomics or metabolomics, emphasizes the complexity of such work; by definition it refers to 'all constituents considered collectively'. Genomics, for instance, is defined as the study of the human genome referring to all the genes of the considered organism and not just a small subset of genes that determines the observed phenotype. Equally, materiomics entails much more than the commonly used approach of piece-wise unravelling the

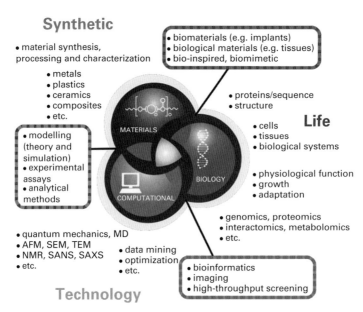

Figure 1.3   **Materiomics – the convergence of disparate fields**. The interface of materials science ('synthetic') and biology ('life') has been successful in the development of biomaterials, but recent technological advances allow for a truly integrated and holistic multidisciplinary approach. While some biological materials have been investigated from a materials science approach, and some material developments have been inspired by Nature, complete understanding requires convergence of each knowledge base and toolset. For example, one direction has been to uncover the functional relationships of biological materials (e.g. physiological function through proteomics attained via bioinformatics) while another direction systematically characterizes the material properties of tissues via modelling and experimental probes common to materials science (e.g. mechanistic interpretations of function derived from molecular simulation). Materiomics lies at the apex of these information streams, attempting to reconcile biological function with material interactions and properties.

properties and behaviour of a material. It entails the study of all possible functionalities and properties. For example, the process of bone tissue growth on a calcium phosphate scaffold under controlled conditions is a materials science and biological problem (albeit nontrivial). Understanding how bone tissue can be grown on *any* arbitrary material platform is a materiomics problem. At the juncture is the emergence of the *materiome*, which can be thought of as the abstract collection of all material behaviours, functions and interactions with all potential material systems and environmental conditions.

Innovation and successful (predictive) biomaterial design involves a rigorous under-standing of the properties and mechanisms of biological matter. Thus, even without the widespread adoption of the term 'materiomics' attempts are currently being made to combine the fields of biology and materials science, resulting in progress in research on complex biological and synthetic material systems. Although biological materials may appear irreducibly complex, researchers in biomaterial synthesis and self-assembly are far from idle. Several spin-off research areas have emerged to satisfy the needs of materials research driven by this new approach. Some of these are biologically 'themed' interdisciplinary research areas, such as bioinformatics, nanobiology or systems biology (see Figure 1.3). Through the merging of technologies, processes and devices, new pathways and opportunities are created that would be inaccessible to any single discipline or knowledge base.

The knowledge and advanced technology acquired over the years by material scientists has allowed the production of large material libraries ('living' and/or synthetic) with diverse chemical properties (28). These include libraries based on block copolymer chemistry (29, 30), or click-chemistry (31–34) or surface topography, as well the protein databank archives (http://www.rcsb.org/). The assembly of such libraries is obviously crucial for progress in the materiomics approach; however, the existence of material *data* should be distinguished from material *knowledge*. While assembly of material libraries is important (and necessary), without associated understanding of material function it is akin to filling a library with books, yet being unable to read a single word. The assembly of materials represents only the first steps in materiomics-based material development, just as determining the genome sequence is the first step in unlocking the power of the genetic code. The analysis of material properties with respect to their biological and functional influence is the variable to address, as inspired by nature, where the structure and function of a system are intimately interlinked. Equally, slight alterations in underlying chemistry of a biological system may have great influence on its resulting functional properties, and may serve to inspire or guide further materials development.

1.5   Conclusion and book outline

The field of materials research for biomedical and clinical applications has witnessed exciting developments over the past several years: a materiomics approach has been undertaken and is forecast to guide the field to progress faster and more efficiently. From the materiomics perspective, biomedical materials research must rely on a holistic approach to investigate biological material systems. As most material properties are strongly dependent on the scale of observation, integration of multiscale experimental and simulation analyses is the key to improve our systematic understanding of how structure and properties are linked. Different scientific fields, with their distinct knowledge and methodologies must converge.

As we have said, at the interface of living and non-living materials, the whole is greater than the sum of its parts. Understanding of such complex systems, therefore, requires more than the summation of disciplinary contributions – fields and techniques must be integrated in a cohesive and synergistic manner. Further streamlining of the process from material banking to assay development, high-content imaging and data mining will ensure that the *materiomics* approach becomes available for the biomaterial research community.

A key challenge is to extend physiochemical metrics, using insights based on the material properties (discussed further in **Chapter 2**) and mechanical function in a biological context, across the molecular, cellular and tissue scales. Seamless integration of synthetic components requires both the development of *de novo* materials (**Chapter 3**) and efficient means to synthesize functional systems (**Chapter 4**). Assessing success requires the integration of advanced biological assays (**Chapter 5**), high-content image processing (**Chapter 6**) and bioinformatics (**Chapter 7**), to evaluate, monitor and predict mechanisms associated with materials and the structures composed of these materials.

Although complete understanding of a material system (biological or synthetic) is theoretically desirable, advances and immediate applications can be developed in a continuously refined and self-regulating cycle. Lacking a complete picture from genetic transcription to tissue function has not impeded advances in and use of the mechanisms and interactions we know (fairly) well (see **Chapter 8** on upscaling, and **Chapter 9** on clinical translation). The study of hierarchical material structures and their effect on molecular and microscopic properties, by making use of structure–process–property relations in a biological context, provides a basis for understanding complex systems by translating material concepts from biology intended for non-biomedical applications (**Chapter 10**). Engineering hubris and ingenuity, combined with clinical need, have laid the groundwork for future refinement. But the inverse problem remains: that is, can we introduce and exploit biological processes (such as healing or growth) seamlessly within a synthetic system, subtly eliminating the distinction between 'material' and 'tissue'? This book is a guide through this materiomics approach to biomaterial research, from material properties to clinical translation.

Further reading

Albright AL, Pollack IF, Adelson PD. *Principles and Practice of Pediatric Neurosurgery* 2nd edn: Thieme; 2008.

Hook AL, Anderson DG, Langer R. High throughput methods applied in biomaterial development and discovery. Biomaterials. 2010;**31**(2): 187–98.

Meekeren JJ. *Observationes Medico-chirurgicae*. Ex Officina Henrici & Vidnae Theodoi Boom; 1682 [In Latin].

Potyrailo R, Rajan K, Stoewe K. Combinatorial and high-throughput screening of materials libraries: review of state of the art. ACS Comb. Sci. 2011;**13** (6):579–633.

Simon CG Jr, Lin-Gibson S. Combinatorial and high-throughput screening of biomaterials. Adv Mater Special Issue: Polymer Science at NIST. 2011;**23**(3) :369–87.

References

1. Langer R, Tirrell DA. Designing materials for biology and medicine. Nature. 2004;**428** (6982):487–92.
2. Burg KJL, Porter S, Kellam JF. Biomaterial developments for bone tissue engineering. Biomaterials. 2000;**21**(23):2347–59.
3. Ma PX. Biomimetic materials for tissue engineering. Adv Drug Deliver Rev. 2008;**60**(2):184–98.
4. Shin H, Jo S, Mikos AG. Biomimetic materials for tissue engineering. Biomaterials. 2003;**24** (24):4353–64.

5. Langer R, Vacanti JP. Tissue engineering. Science. 1993;**260**(5110):920–6.
6. de Bruijn JD, Yuan HP, Fernandes H *et al*. Osteoinductive ceramics as a synthetic alternative to autologous bone grafting. Proc Natl Acad Sci USA. 2010;**107**(31):13614–19.
7. Neuss S, Apel C, Buttler P *et al*. Assessment of stem cell/biomaterial combinations for stem cell-based tissue engineering. Biomaterials. 2008;**29**(3):302–13.
8. Fratzl P, Weinkamer R. Nature's hierarchical materials. Prog Mater Sci. 2007;**52**(8):1263–334.
9. Buehler MJ, Yung YC. Deformation and failure of protein materials in physiologically extreme conditions and disease. Nat Mater. 2009;**8**(3):175–88.
10. Csete ME, Doyle JC. Reverse engineering of biological complexity. Science. 2002;**295** (5560):1664–9.
11. Unadkat HV, Hulsman M, Cornelissen K *et al*. An algorithm-based topographical biomaterials library to instruct cell fate. Proc Natl Acad Sci USA. 2011;10.1073/pnas.1109861108.
12. Knowles TPJ, Buehler MJ. Nanomechanics of functional and pathological amyloid materials. Nat Nanotechnol. 2011;**6**(8):469–79.
13. Buehler MJ. Tu(r)ning weakness to strength. Nano Today. 2010;**5**(5):379–83.
14. Cranford S, Buehler MJ. Materiomics: biological protein materials, from nano to macro. Nanotechnol Sci Appl. 2010;**3**(1):127–48.
15. Ackbarow T, Buehler MJ. Hierarchical coexistence of universality and diversity controls robustness and multi-functionality in protein materials. J Comput Theor Nanos. 2008;**5** (7):1193–204.
16. Buehler MJ. Nanomaterials: Strength in numbers. Nat Nanotechnol. 2010;**5**(3):172–4.
17. Titz B, Schlesner M, Uetz P. What do we learn from high-throughput protein interaction data? Expert Rev Proteomic. 2004;**1**(1):111–21.
18. Govorun VM, Archakov AI. Proteomic technologies in modern biomedical science. Biochem Moscow. 2002;**67**(10):1109–23.
19. Pandey A, Mann M. Proteomics to study genes and genomes. Nature. 2000;**405**(6788):837–46.
20. Aizenberg J, Fratzl P. Biological and biomimetic materials. Adv Mater. 2009;**21**(4):387–8.
21. Vincent JFV, Bogatyreva OA, Bogatyrev NR, Bowyer A, Pahl AK. Biomimetics: its practice and theory. J R Soc Interface. 2006;**3**(9):471–82.
22. Webster DC. Combinatorial and high-throughput methods in macromolecular materials research and development. Macromol Chem Phys. 2008;**209**(3):237–46.
23. Rademann J, Jung G. Drug discovery: Integrating combinatorial synthesis and bioassays. Science. 2000;**287**(5460):1947–8.
24. Westerhoff HV, Bruggeman FJ. The nature of systems biology. Trends Microbiol. 2007;**15** (1):45–50.
25. Venter JC. Multiple personal genomes await. Nature. 2010;**464**(7289):676–7.
26. Rogers YH, Venter JC. Genomics: Massively parallel sequencing. Nature. 2005;**437** (7057):326–7.
27. Kohn J, Welsh WJ, Knight D. A new approach to the rationale discovery of polymeric biomaterials. Biomaterials. 2007;**28**(29):4171–7.
28. Xiang XD, Sun XD, Briceno G *et al*. A combinatorial approach to materials discovery. Science. 1995;**268**(5218):1738–40.
29. Langer R, Anderson DG, Levenberg S. Nanolitre-scale synthesis of arrayed biomaterials and application to human embryonic stem cells. Nat Biotechnol. 2004;**22**(7):863–6.
30. Oreffo ROC, Tare RS, Khan F *et al*. A microarray approach to the identification of polyurethanes for the isolation of human skeletal progenitor cells and augmentation of skeletal cell growth. Biomaterials. 2009;**30**(6):1045–55.

31. Capito RM, Azevedo HS, Velichko YS, Mata A, Stupp SI. Self-assembly of large and small molecules into hierarchically ordered sacs and membranes. Science. 2008;**319**(5871):1812–16.

32. Silva GA, Czeisler C, Niece KL *et al*. Selective differentiation of neural progenitor cells by high-epitope density nanofibers. Science. 2004;**303**(5662):1352–5.

33. Stupp SI, Braun PV. Molecular manipulation of microstructures: Biomaterials, ceramics, and semiconductors. Science. 1997;**277**(5330):1242–8.

34. Meijer EW, Dankers PYW, Harmsen MC, Brouwer LA, Van Luyn MJA. A modular and supramolecular approach to bioactive scaffolds for tissue engineering. Nat Mater. 2005;**4**(7):568–74.

2 Physico-chemical material properties and analysis techniques relevant in high-throughput biomaterials research

Erwin Zant, Mirjam J. Bosman and Dirk W. Grijpma

Scope

High-throughput and combinatorial research on biomaterials aims at the rapid development of new materials and the establishment of structure–function relationships. Therefore, knowledge of the chemistry of each material and its impact on physical properties is essential to understand the effect on its function as a biomaterial. A tutorial on basic physical and chemical properties of (polymeric) materials highlights these features and can be found in the first part of this chapter. The second part gives an overview of relevant techniques that can be used to screen these material properties in high throughput. In addition, several examples are described in which these methods are used to develop structure–function relationships between material properties and biological performance.

2.1 Basic principles: physical and chemical properties of polymeric biomaterials

Chemistry is a constant factor from which the performance of most (polymeric) bio-materials can be predicted, but this extrapolation becomes less obvious when numerous materials are mixed in huge combinatorial libraries. Therefore, researchers are becoming increasingly involved in high-throughput material research when successful correlations between biological performance and physico-chemical material properties are to be made. This accelerating trend can be extracted from many studies where high-throughput technologies are successfully applied to measure physical properties and biological performance of many different polymeric biomaterials. Physical properties such as hardness, topography and hydrophilicity are known to be important parameters in the biological evaluation of materials, because they allow or block the adhesion of biological compounds which is required to allow cell-spreading, migration, proliferation and differentiation. These properties are naturally different for every material or combination of

Materiomics: High-Throughput Screening of Biomaterial Properties, ed. Jan de Boer and Clemens van Blitterswijk. Published by Cambridge University Press. © Cambridge University Press 2013.

materials, and relate primarily to the variable properties on the chemical level (molecular structure, functional groups and degradation). Therefore, the chemistry of a biomaterial directly contributes to its interaction with biological environments. Combinatorial chemistry and high-throughput analysis is being developed to correlate these material properties to the biological outcome and eventually find the ultimate biomaterial.

2.1.1   Bulk characteristics

Chemical properties

The properties of polymeric biomaterials on the chemical level deal with the information on the structural units, namely what type of monomer comprises the polymer chain and whether one or more than one type of monomer (copolymer) is used. Two types of structural units can be recognized: functional and non-functional units. The non-functional units comprise the hydrocarbon building blocks of the polymer chain, methylene ($-CH_2-$) and phenylene ($-C_6H_4-$) for example; in both groups the hydrogen atoms can be interchanged by other functionalities or groups. The functional structural units can originate from condensation reactions of functional monomers (functionalized by $-OH$, $-NH_2$, $-COOH$, $-COCl$, etc.). These functional units determine the characteristic names of polymer families, such as polycarbonates, -esters, -amides or -urethanes. These groups affect the thermal and mechanical properties of polymers, which largely determine their field of application. Furthermore, functional groups determine whether the polymer is prone to be affected by degradation processes such as hydrolysis. This degradation process can be very useful in some polymers, making them biodegradable and therefore applicable in medical implants, drug delivery devices (1) and tissue engineering scaffolds (2, 3).

Biodegradation

The development of biodegradable polymers has attracted a lot of interest from clinicians, because it can be very useful when polymers are used as artificial implants and a second round of surgery is not needed. Moreover, this class of materials can lead to better recovery of the tissue and therefore is useful in tissue engineering where tissue repair or remodelling is the goal (4). In brief, biodegradation is the chemical degradation, or backbone breakdown, of a polymer chain due to hydrolytic or enzymatic activity (5). Therefore, only polymers with labile structural units such as esters, anhydrides, carbonates, orthoesters and amides (Figure 2.1) suit the application.

Naturally derived biodegradable polymers such as collagen, glycosaminoglycans, chitosan and polyhydroxyalkanoates, representing polyamides (polypeptides), polysaccharides and polyesters respectively, have excellent biological properties and are biocompatible (6). Their synthetic counterparts possess large-scale reproducibility and can be processed into tissue engineering products in which the mechanical properties and degradation time can be controlled. Examples of synthetic polyesters are poly(glycolic acid) (PGA) and poly(lactic acid) (PLA), which are the most widely used synthetic degradable polymers in medicine. The relative hydrophilicity of PGA makes it very fast to biodegrade because the absorption of water accelerates hydrolysis, but copolymerization with the more hydrophobic polymer PLA makes the material suitable for a

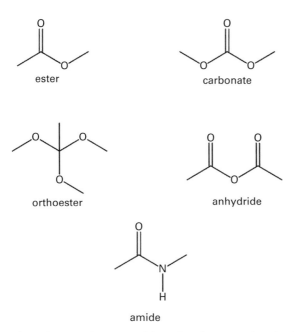

Figure 2.1   Shown here are the structural formulas of common functional groups which are vulnerable in biological environments and make the biodegradation of polymers possible.

wider range of applications. Poly(ε-caprolactone) (PCL) characterizes another polyester with somewhat slower degradation than PGA and PLA, which favours its use in long-term, implantable systems. In summary, biodegradation influences the mechanical and biological properties of polymers over time and is therefore an important factor in the development of new biomaterials.

Thermal properties

The thermal properties of polymeric biomaterials depend on two morphological domains in the bulk of the material: the amorphous phase and the crystalline regions. The distribution of amorphous polymer chains in the matrix is completely random in the amorphous state (glass) and frozen in position. This phase contributes to the thermal transition from glass to rubber, the glass transition temperature (T_g), where the molecular motion in the amorphous phase of the polymer increases and the polymer becomes rubbery. On increasing the temperature of the polymer even further, the amorphous material starts to decrease in viscosity and finally flows. Whereas amorphous polymer domains are disordered structures, crystalline regions are areas where the polymer chains are highly ordered. These become mobile above the melting point (T_m) of the polymer, which is much higher than the glass transition temperature of most polymers.

While increasing the temperature of semi-crystalline polymer materials, one can identify a glass phase, a rubber phase and a molten phase. Transitions between the phases lead to different mechanical properties: in the glass state (below the polymer's glass transition temperature), the material is rigid and difficult to deform; the rubber plateau is the region above the glass transition temperature where the material is flexible and easier to deform. Poly(trimethylene carbonate) (PTMC) is such an amorphous polymer, which has a T_g around −15 °C and has no melting temperature; therefore it shows a tensile modulus (an indication of stiffness) of around 6–7 MPa at room temperature (7). However, semi-crystalline polymers are able to maintain higher stiffness above the glass transition temperature, since the crystalline regions act as physical entanglements between the mobile rubber domains. This behaviour can be observed in the biodegradable polyester PCL, which has a T_g of −60 °C, a T_m of 59 °C and a tensile modulus of 400 MPa at room temperature (8), which is much higher than that for PTMC. Crystallinity in the bulk also affects the morphology at the surface by increasing surface roughness, and hence can have a significant effect on biological interactions. Cells respond to this phenomenon by decreasing their proliferation rate with increasing surface roughness (9).

Mechanical properties
Static measurements

The mechanical properties of biomaterials are of great importance, in particular when they are used as structural components. The most often determined mechanical parameter in high throughput is the elasticity modulus, which is the ratio between the stress and the applied deformation (2). Three different modes of static deformation exist: tensile, compression and shear deformation, from which the E-, K- and G-modulus are the derived moduli, respectively. The E-modulus or Young's modulus describes the material stiffness at small strains or the resistance of the material to reversible deformation and is defined as the initial slope of a stress–strain diagram obtained during a uniaxial tensile test (see Figure 2.2).

In polymers, the elastic modulus mostly origins from secondary (interchain) interactions, mainly van der Waals interactions (10), and the resulting E-modulus of

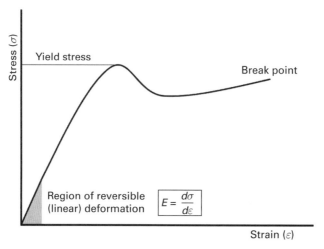

Figure 2.2   Tensile stress (σ) versus tensile strain (ε) of ideal thermoplastic polymers. The slope of the tensile curve in the linear deformation is a measure of the modulus or stiffness.

glassy polymers typically ranges from 2.5 to 3.5 GPa. Yield stress and elongation at break are two other important parameters which define the material as respectively strong or weak, and brittle or tough.

Yielding, strain softening and strain hardening are processes that often occur when stress is applied on a polymer. Yield is defined as the stress at which the polymer deforms plastically and deformation is irreversible. Yielding is often followed by strain softening as the engineering deformation stress applied to the polymer decreases; strain hardening results from the resistance to deformation that often occurs subsequently. The mechanical performance of a polymer depends on its glass transition temperature and its molecular weight between entanglements (which determines the entanglement density) (11).

Dynamic measurements

Polymers are viscoelastic materials, which means that they show both solid (elastic) and fluid (flow) behaviour (12). The elasticity and flow of polymers are time-dependent and can be visualized using dynamic mechanical analysis (DMA). In DMA a small force is applied on the polymer sample so that the sample is always within the reversible/elastic region of its stress–strain curve, and the resulting strain is determined (13). The applied force is sinusoidal; hence the modulus can be expressed as an in-phase component, the storage modulus, and an out-of-phase component, the loss modulus. During these dynamic experiments, part of the material behaves elastically (this comprises the in-phase component) and another part in a plastic manner (which causes the out-of-phase response). When the material is completely viscous the phase-shift is 90°; when the material is completely elastic the phase shift is 0°.

To be practical in the field of (polymeric) biomaterials, DMA can be used to test materials over a broad temperature range to determine thermal transitions such as the glass transition temperature.

Hardness measurements

The material hardness is a measure of the resistance to local deformation of a material (14, 15). It depends mainly on the elastic modulus of the bulk. The parameter is determined at the surface and offers a straightforward and fast evaluation of the bulk mechanical properties of a material (16, 17). Three methods are available to measure hardness: scratch hardness, rebound hardness and indentation hardness. Among these three, indentation (or depth-sensing indentation) is the most commonly used technique for the analyses of polymer mechanical properties. Different polymers can have very different material hardness values, ranging from soft gel-like materials to glasses, and different indenter geometries exist, varying from diamond pins to soft rubber spheres. Standardized methods such as Shore or Rockwell hardness determination methods use fixed forces and geometries for each scale type (Shore A uses, for instance, a 35° hardened steel cone with diameter 1.40 mm, height 2.54 mm and force 8 N), which limit their applicability in high-throughput screening. Instrumented micro-, ultramicro- or nanoindentation simultaneously measures the load to indent and the displacement of the probe into the surface of the material. From these two parameters, the mechanical properties of a broad range of materials can be evaluated (18).

Box 2.1 Classic experiment

Cited more than 1700 times (Web of Knowledge, April 2012), the study performed by Engler *et al.* in 2006 was and still is very meaningful in the biomaterial research field (49). The research reinforced the fundaments of the correlations between physical properties of substrate materials and stem cell commitment. This paper showed that mesenchymal stem cells can differentiate into particular cell lineages just by sensing the stiffness of the substrate. Substrates with various elastic modulus values were obtained by the synthesis of poly(acrylamide) gels in which the concentration of bis-acrylamide determined the crosslink density, hence the rigidity upon swelling in water. The *E*-modulus was determined by atomic force microscopy (AFM) and varied from 0.1 kPa to 40 kPa.

Characterizing the cells by observing cell shape, measuring gene expression and staining specific proteins showed that mesenchymal stem cells react to the stiffness of the substrate by differentiation into different cell lineages. The stiffness of the tissue to which the cells differentiated corresponded to the stiffness of the substrate on which the cells were cultured. This research provided a breakthrough in the field of both high-throughput screening of material properties and the correlation between substrate stiffness and stem cell fate.

2.1.2   Surface characteristics

Surface chemistry has an important effect on cell behaviour

The chemical nature of surfaces plays an important role in the performance of polymeric biomaterials. For instance, polymer chemistry has a significant effect on the behaviour of cells at the surface, as was described by Folkman and Moscona in 1978 (19). The study concerned the seeding of cells on tissue-cultured polystyrene surfaces with coatings varying in concentrations of the hydrophilic polymer poly(2-hydroxyethyl methacrylate) (pHEMA). Cell spreading was reflected by the average cell height and was found to be higher with low amounts of pHEMA. The spreading correlated with the rate of cell growth. These experiments showed that the chemical character of polymer surfaces can have important consequences for cell shape and cell proliferation. As a consequence, many researchers started to design more surface chemistries to correlate with cell function (20–23).

Surface modifications

Chemical surface modification or the covalent immobilization of bioactive compounds onto functionalized polymer surfaces plays a key role in determining interfacial interactions between the polymeric material and biological media (such as protein solutions, cells or tissues) (4, 24). Polystyrene substrates for tissue engineering are, for instance, treated by glow discharge or exposure to chemicals such as sulphuric acid, to increase the number of charged groups at the surface, which improves cell adhesion and proliferation of many types of cells. Modifying polymeric surfaces by introducing functional groups is a chemical strategy to tune protein and cell adhesion. Several surface modification techniques have been developed to improve wetting, adhesion and printing of polymer surfaces by introducing a variety of polar groups such as amines, carboxylic acids, thiols and hydroxyls, or by the immobilization of proteins and peptides (25).

Another method that is widely applied is polymer grafting: a polymer is 'grown' on the surface of another polymer. In this way the desired properties of the two polymers can be combined. For instance, a polymer with good cell adhesion properties can be grafted onto a polymer with desired mechanical and thermal properties, without significantly influencing these properties. By tuning the density and molecular weight of the grafted polymer, the properties can be fine-tuned for specific applications (26–28). Polymer–cell interactions can also be tuned by incorporating specific bioactive molecules on the polymer surface, which either improve cell adhesion or prevent unspecific and/or unfavourable reactions. Heparin-based coatings are widely used for this purpose, as this molecule increases the biocompatibility of polymer surfaces *in vitro* as well as *in vivo*. Since most of the cell functions are mediated by protein–protein interactions, proteins are also often coupled to polymer surfaces in order to increase the biocompatibility. Extracellular proteins, such as collagen, elastin, fibrin, albumin and immunoglobulins, are the most commonly used for this, since they play a major role in cell adhesion, spreading and growth regulation (25, 29). In addition to proteins, peptides may be applied to enhance the stability of the bioactive molecules immobilized on the polymer surface. The arginine-glycine-aspartic acid (RGD) peptide is mostly used for this purpose. Growth factors can also be incorporated on the polymer surface to stimulate cell growth, proliferation and differentiation (25).

Wettability

Hydrophilicity is probably the most important parameter in cell-related studies (30). Wettable or hydrophilic surfaces have the tendency to interact with or be dissolved by water molecules. Cell adherence to surfaces is very dependent on this parameter, since cell adhesion has been found to be optimal at intermediate hydrophilicity (30, 31), whereas proliferation increases with increasing hydrophobicity (32, 33). Several treatments and material bulk properties influence the wettability of surfaces. For instance, increasing the number of charged groups on the surface affects wettability, because hydrogen bonding with the water molecules is enhanced and the droplet spreads along the hydrophilic surface, resulting in a lower contact angle (25). Cells respond to this by an increase in adhesion (30, 31). Increased polymer crystallinity leads to increased surface roughness on a nanometre scale, which optimizes hydrophilicity and affects cell proliferation (9). Cell adhesion is enhanced when surfaces are pretreated with proteins, and protein adsorption is again dependent on hydrophilicity. Therefore cell adhesion is influenced by two parameters: hydrophilicity and protein adsorption (30). The most conventional method of assessing wettability is to perform water contact angle measurements to measure how a water droplet spreads on a surface.

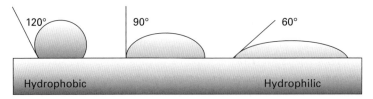

Figure 2.3   Water contact angle measurements. Hydrophobic surfaces show higher contact angles.

As shown in Figure 2.3, the lower the contact angle, the more hydrophilic is the surface (34).

Topography

The surface morphology or topography of materials has a strong influence on cell behaviour: the response of cells (e.g. adhesion and proliferation) on patterned surfaces (nanometre to micrometre scale) is different from the behaviour on smooth surfaces. For instance, Riehle and co-workers have demonstrated that highly ordered nanotopographies result in negligible to low cellular adhesion and osteoblastic differentiation, whereas mesenchymal stem cells on random nanotopographies exhibited a more osteoblastic morphology (35). Research in the laboratory of Simon and co-workers showed that with a gradient in polymer crystallinity in poly(L-lactic acid) films, the surface roughness was affected and cell proliferation was found to be inversely correlated with the surface roughness (9). Topography was also shown to be a driving force in a study that evaluated the response of MC3T3-E1 cells to dimethacrylate composites. The biomaterials used varied in filler content, degree of methacrylate conversion and surface roughness, hence a combinatorial testing platform was developed. Overall, the cell response was found to depend on multiple material properties. For instance, cell viability was highest at higher degrees of conversion, a smoother surface and more hydrophilic regions, and was only mildly affected by filler type and content. At lower degrees of conversion the surface of the materials was rougher and more hydrophobic, and cells did not spread as well as on smooth surfaces (36).

2.2   Relation to materiomics: techniques that allow high-throughput material characterization

Conventional material characterization methods often lack the option of repeating measurements many times in a short time period. Nevertheless, other methods have been developed or modified to allow rapid material characterization. Relevant and frequently applied high-throughput techniques that are used to analyse polymer properties such as topography, hydrophilicity, stiffness and chemistry are atomic force microscopy (AFM), X-ray photoelectron spectroscopy (XPS), time-of-flight secondary ion mass spectrometry (ToF-SIMS), Fourier transform infrared (FTIR)/Raman spectroscopy, nanoindentation and water contact angle (WCA) measurement.

2.2.1   Bulk characterization in high throughput

High-throughput bulk analysis of chemical properties
Fourier transform infrared (FTIR)/Raman spectroscopy

FTIR/Raman microspectroscopy provides an approach to determine the bulk chemical functionalities of materials (25, 32). On interaction with the electromagnetic waves, chemical bonds stretch, contract and bend, causing the material to absorb infrared radiation of a defined wavenumber. The output plots the absorption as a function of wavenumber, where the wavenumber (which in spectroscopy is used as a unit of energy) identifies a specific chemical bond. Because it can provide the chemical information in seconds, FTIR/ Raman spectroscopy is widely used in high-throughput analysis of combinatorial and gradient material arrays. It was, for instance, applied in a study of a 2D gradient specimen, where both monomer composition and degree of conversion were analysed simultaneously, providing a qualitative analysis of the polymer chemistry (37). High-throughput application of FTIR spectroscopy also proved very practical in the characterization of biodegradable polyanhydride compositions produced in a gradient-like library, where a correlation with phase-behaviour results from optical and atomic force microscopy was performed to provide information about polymer composition, upper critical solution temperature and surface roughness (38). Hence, FTIR can be a helpful tool to analyse libraries of materials where the composition at different positions in the library is unknown.

High-throughput bulk analysis of mechanical properties
Nanoindentation

The technique most often used to assess the mechanical properties of materials is tensile testing. However, tensile testing is not well suited to high-throughput techniques, so other methods for the high-throughput analysis of mechanical properties have been developed. One of the most common is nanoindentation, which refers basically to a technique where a hardness test is performed at the nanometre scale, so only small amounts of material are required to quantify the mechanical properties (14, 17, 18). The method is also known as depth-sensing indentation, where the name already denotes the principle of operation. Both the load and the displacement of the probe into the polymer are measured simultaneously.

As an example, a schematic representation of a typical dataset obtained with a Berkovich indenter is presented in Figure 2.4. The parameter P designates the load and h the displacement relative to the initial non-deformed surface. The material response of polymers to deformation is assumed to be elastic and plastic in nature during indentation, whereas during unloading of the indenter only the elastic displacements are recovered. Therefore, the unloading curve is useful in the analysis (17). In conventional tensile tests, the stiffness or elastic modulus is defined as the initial slope of the stress–strain diagram. The stiffness of the material S during indentation is defined as the slope of the upper portion of the unloading curve (17).

As said earlier, Figure 2.4 shows the response of a conical diamond-shaped or Berkovich indenter, which has been used often in this particular application. Results obtained with such devices need to be used with caution, as undesired effects can occur, such as material pile-up along the indenter walls which increases the contact area between the probe and the material. Hence other indenter tips have been used for nanoindentation (18). For instance, the flat punch probe can be useful, as the surface of indentation remains constant, but it is sensitive to misalignment and therefore not very useful while scanning a large library with numerous height differences. A spherical indenter should prevent most misalignment problems, and this has been used in the high-throughput context as well (39).

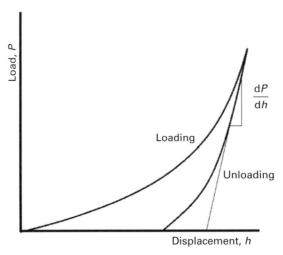

Figure 2.4   Schematic illustration of a load (*P*) versus displacement (*h*) indentation showing important measured parameters, which allow the calculation of the material stiffness (*S*) during the unloading process.

Box 2.2 Classic experiment

Recently, collaboration between several groups resulted in a publication where the usability of high-throughput research was assessed (48). This investigation represents the study of cell colony formation on microarrays comprising different polymeric biomaterials with the use of numerous high-throughput techniques such as picolitre water contact angle WCA (wettability) measurements, AFM (surface topography), ToF-SIMS (surface chemistry), nanoindentation (rigidity) and laser-scanning cytometry (cell quantification). The microarrays were prepared from 22 different acrylate monomers containing several functional structural units with varying hydrophobicities and chain lengths. By combining the monomers in different ratios, the microarrays comprised 496 different materials. These mixtures were robotically deposited on pHEMA-treated glass slides and then polymerized with ultraviolet (UV) light. The specimens on the glass slides were evaluated for their material properties such as elasticity modulus, wettability and surface roughness. Binning the results into heat maps yielded a good presentation of the physical properties of the materials and the corresponding effect on colony formation. Using these results, the researchers were able to correlate the different parameters to cell colony formation.

Surface characterization in high throughput

High-throughput surface analysis of chemical properties
X-ray photoelectron spectroscopy
XPS, or electron spectroscopy for chemical analysis (ESCA), determines the atomic composition of the surface of a solid to a depth of several nanometres. Upon exposure to X-ray photons, a surface ejects photoelectrons, and these binding energies can be compared with reference values to identify the element and its oxidation state. The resulting spectrum is a plot of intensity versus binding energy. The intensity of the ejected photoelectrons relates directly to the atomic distribution of the material surface and can therefore be used to quantify percentage atomic composition and stoichiometric ratios (25). Since XPS is a rapid technique, it is suitable for high-throughput analysis. For instance, a library of 576 different novel surface chemistries was analysed with XPS offering a fast method of producing a library of chemical data (40–42). Studies performed in the group of Voelcker also resulted in the successful application of XPS. They generated multifunctional surface chemistries using poly (ethylene glycol)-methacrylates as a non-cell-adhering background and glycidyl meth-acrylate linkers to couple a variety of biologically active molecules (43).

Time-of-flight secondary ion mass spectrometry
ToF-SIMS is a mass spectroscopy technique that is commonly used as a complementary technique with XPS, since ToF-SIMS does not provide sufficiently quantitative results. ToF-SIMS is more sensitive than XPS; it makes use of primary ions which are sputtered on the sample surface causing secondary ions to eject from the surface. The mass of these secondary ions is determined by measuring the time needed for them to move from the surface of the material to the detector. Hence, detailed information about the type and quantity of ionizable chemical groups of a surface to a depth of a few nanometres can be obtained (25). The resulting spectrum depicts signal intensity versus mass-to-charge ratio and can be used to determine relative intensities of chemical species. For high-throughput analysis, ToF-SIMS spectra of all samples of the library are acquired. Ion distribution images of the entire library then allow the rapid screening of the presence of certain ions (40–42).

High-throughput surface analysis of physical properties
Water contact angle
WCA allows the assessment of surface hydrophilicity by measuring how much a droplet of water spreads on a surface. The lower the contact angle, the more hydrophilic the surface is (25). However, the method is generally limited to macroscopic measurements because the base diameter of the droplet is usually greater than 1 mm. To allow WCA measurements in high throughput, piezoelectric dispensers producing picolitre sized drop-lets with micrometre precision were developed and applied (40, 44). A piezo-dosing unit, similar to those used in inkjet printers, dispenses picolitre volumes of water onto the material, yielding drops with a diameter of approximately 70 μm (44). For high-throughput purposes, sample positioning and data acquisition can be automated using

dual camera systems (one camera records a side profile of the drop, the other provides an overhead view) to ensure deposition on the centre of each sample (40).

Atomic force microscopy

In AFM, a cantilever with a tip moves over the surface. As the tip interacts with the surface it screens the surface of the polymer. The change in surface height is then measured by the location of a reflected laser beam in a photodetector and the surface topography is mapped. From this, the surface roughness can be determined. In tapping-mode AFM, any friction between the tip and the surface that could distort the obtained image is avoided. AFM has been used very often in high-throughput analyses of surfaces, because it very rapidly generates topography maps with nanometre resolution.

Polymer bulk crystallinity results in topographical changes at the surface of materials, as was discovered using AFM in the case of phase separation during cooling of two polymers (36) and different annealing temperatures for semi-crystalline polymers (9). The AFM apparatus was very useful in both cases to correlate surface roughness with bulk crystallinity, morphology and biological interactions. As described by Yang *et al.* (41), surface smoothness was essential to eliminate any influence of the topography on cellular response. AFM analysis revealed that the majority of the samples had a very low surface roughness (below 5 nm). Without surface roughness as a confounding factor, a better correlation between surface chemical functionality and cell–material interaction could be made (41). Furthermore, although it has not often been applied in high-throughput mechanical testing, AFM has proved very useful in the nanomechanical screening of ultrathin films (1–100 nm) (45).

2.3   Future perspectives

The development of novel fast synthesis and analysis techniques will become more and more vital when huge numbers of materials with varying chemistries and physics are concerned. Nowadays, several production techniques, as shown schematically in Figure 2.5, already exist to aid in the fabrication of enormous material libraries within several minutes and with minimal material usage. Automatic liquid handling systems have been developed to decrease monomer mixing-time, and inkjets or robot-driven printing heads have proven useful in the production of material libraries in microarray format.

Polymer grafting or imprinting by photoinduced graft polymerization can be regarded as one of the promising innovations in high-throughput synthesis. This approach was used by the Belfort group for the high-throughput grafting of glycidyl methacrylate and amine compounds on poly(ether sulphone) (PES) in developing low fouling membranes (46). As well as the non-adhesiveness to proteins, glycidyl methacrylate comprises an epoxy moiety which can react with other functional groups by ring opening and poly-condensation. The PES material readily produces radicals upon ultraviolet irradiation, which can subsequently initiate the radical polymerization of the glycidyl methacrylate. The chemistry is shown schematically in Figure 2.6. Using this simple chemistry, the

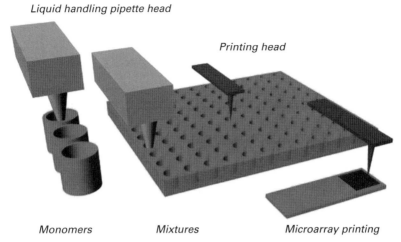

Schematic presentation of the rapid fabrication of microarrays material libraries on glass slides using liquid handling pipette systems or contact printing heads and a plate containing mixtures of various starting materials or monomers.

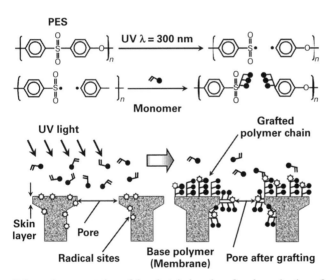

Figure 2.6 Schematic presentation of the photoinduced graft polymerization of monomers containing double bonds (like glycidyl methacrylate) on PES surfaces. Ultraviolet light causes chemical cleavage of the PES, yielding radicals which can initiate the polymerization of (meth)acrylates from the surface. Reprinted (adapted) with permission from (47). Copyright (2007) American Chemical Society.

group was able to produce numerous different membranes by coupling of the glycidyl polymer to 25 different amine monomers.

Belfort developed this technique together with Anderson, and, with colleagues, described the modification of PES surfaces with 66 monomers (47). The fact that the photo-induced graft polymerization was performed in a 96-well format allowed the evaluation of many monomers at the same time, so a high-throughput platform was produced to develop new membranes. Using this chemistry and high-throughput platform, fast polymer synthesis and surface modifications of other biomaterials can also be developed.

The publication discussed in Box 2.2 shows an example of microarray printing (48). Here, high-throughput screening successfully correlated the physico-chemical properties of 496 different materials with possible biological interactions. The starting materials of choice were 16 acrylate monomers with different chemical structures which were copolymerized with 6 other acrylate monomers in 6 ratios (100:0, 90:10, 85:15, 80:20, 75:25, 70:30) yielding 480 mixtures and 16 homopolymeric networks. However, more combinations could be made if the 16 acrylate monomers were also to be mixed with each other, and even more in a combinatorial approach where every monomer was mixed with every other. The latter approach would yield $2^{22} = 4\,194\,304$ different materials. Acquiring all the physical and chemical data from this number of materials is a challenge, and generating new insights will be even harder for any researcher. Therefore, the future of high-throughput material research embraces both the development of rapid production and analysis facilities and the development of improved bioinformatics to handle enormous datasets.

2.4   Snapshot summary

- Chemistry is a constant factor from which the performance of most polymeric biomaterials can be predicted. The properties at the chemical level depend on the structural units, namely, what type of monomer comprises the chain and whether one or more than one type of monomer (copolymer) is used.
- Physical properties are different for each material or combination of materials and relate primarily to properties at the chemical level (molecular structure, functional groups and degradability).
- The interaction of cells with materials depends on physical properties such as hardness, topography and hydrophilicity because these properties allow or prevent the adhesion of biological compounds which are required to allow cell-spreading, migration, proliferation and differentiation.
- To overcome the difficulties related to exponentially growing numbers of materials, high-throughput methods were developed.
- Automated pipettes, printers and analysis techniques facilitate the high-throughput characterization of large material libraries.
- The development of new biomaterials using high-throughput techniques involves both the correlation of physico-chemical properties with cell–material interactions and finding (unexpected) hits where a material induces a desired response.
- The production of material libraries using novel polymerization routes as shown in Section 2.3 will become very important.
- The rapid development of data handling systems and effective informatics will become essential to handle fast-growing datasets.

Further reading

Hook AL, Thissen H, Voelcker NH. Surface plasmon resonance imaging of polymer microarrays to study protein–polymer interactions in high throughput. Langmuir. 2009;**25**(16):9173–81.

Parekh SH, Chatterjee K, Lin-Gibson S *et al*. Modulus-driven differentiation of marrow stromal cells in 3D scaffolds that is independent of myosin-based cytoskeletal tension. Biomaterials. 2011;**32**(9):2256–64.

Kranenburg JM, Tweedie CA, van Vliet KJ, Schubert US, Challenges and progress in high-throughput screening of polymer mechanical properties by indentation. Adv Mater. 2009;**21**:3551–61.

Alexander MR, Taylor M, Urquhart AJ, Zelzer M, Davies MC. Picoliter water contact angle measurement on polymers. Langmuir. 2007;**23**(13):6875–8.

References

1. Uhrich KE, Cannizzaro SM, Langer RS, Shakesheff KM. Polymeric systems for controlled drug release. Chem Rev. 1999;**99**(11):3181–98.
2. Van Krevelen DW, Te Nijenhuis K. *Properties of Polymers* (4th edn). Amsterdam: Elsevier; 2009.
3. Flory PJ. *Principles of Polymer Chemistry*: Cornell University Press; 1953.
4. Lanza R, Langer R, Vacanti J. *Principles of Tissue Engineering* 3rd edn: Elsevier; 2007.
5. Buschow KHJ, Cahn RW, Flemings MC *et al*. *Encyclopedia of Materials: Science and Technology*: Elsevier; 2001.
6. Kim B-S, Park I-K, Hoshiba T *et al*. Design of artificial extracellular matrices for tissue engineering. Prog Polym Sci. 2011;**36**(2):238–68.
7. Pego AP, Grijpma DW, Feijen J. Enhanced mechanical properties of 1,3-trimethylene carbonate polymers and networks. Polymer. 2003;**44**(21):6495–504.
8. Pego AP, Poot AA, Grijpma DW, Feijen J. Copolymers of trimethylene carbonate and epsilon-caprolactone for porous nerve guides: Synthesis and properties. J Biomat Sci Polym E. 2001;**12**(1):35–53.
9. Washburn NR, Yamada KM, Simon Jr CG, Kennedy SB, Amis EJ. High-throughput investigation of osteoblast response to polymer crystallinity: Influence of nanometer-scale roughness on proliferation. Biomaterials. 2004;**25**(7–8):1215–24.
10. Meijer HEH, Govaert LE. Mechanical performance of polymer systems: The relation between structure and properties. Prog Polym Sci. 2005;**30**(8–9):915–38.
11. Meijer HEH, Govaert LE. Multi-scale analysis of mechanical properties of amorphous polymer systems. Macromol Chem Phys. 2003;**204**(2):274–88.
12. Reiner M. The Deborah number. Phys Today. 1964;**17**(1):62.
13. Eisele U. *Introduction to Polymer Physics* 1st edn: Springer-Verlag; 1990.
14. Tweedie CA, Anderson DG, Langer R, Van Vliet KJ. Combinatorial material mechanics: High-throughput polymer synthesis and nanomechanical screening. Adv Mater. 2005;**17** (21):2599–+.
15. Sundararajan G, Roy M. Hardness testing. In: Robert WC, Merton CF, Bernard I *et al*., eds. *Encyclopedia of Materials: Science and Technology*: Elsevier; 2001. p. 3728–36.
16. Oliver WC, Pharr GM. An improved technique for determining hardness and elastic modulus using load and displacement sensing indentation experiments. J Mater Res. 1992;**7**(6):1564–83.

17. Oliver WC, Pharr GM. Measurement of hardness and elastic modulus by instrumented indentation: Advances in understanding and refinements to methodology. J Mater Res. 2004;**19**(1):3–20.

18. Kranenburg JM, Tweedie CA, van Vliet KJ, Schubert US. Challenges and progress in high-throughput screening of polymer mechanical properties by indentation. Adv Mater. 2009;**21**(35):3551–61.

19. Folkman J, Moscona A. Role of cell-shape in growth control. Nature. 1978;**273**(5661):345–9.

20. Anderson DG, Putnam D, Lavik EB, Mahmood TA, Langer R. Biomaterial microarrays: rapid, microscale screening of polymer–cell interaction. Biomaterials. 2005;**26**(23):4892–7.

21. Brocchini S, James K, Tangpasuthadol V, Kohn J. Structure–property correlations in a combinatorial library of degradable biomaterials. J Biomed Mater Res. 1998;**42**(1):66–75.

22. Hansen A, McMillan L, Morrison A, Petrik J, Bradley M. Polymers for the rapid and effective activation and aggregation of platelets. Biomaterials. 2011;**32**(29):7034–41.

23. Khan F, Tare RS, Kanczler JM, Oreffo ROC, Bradley M. Strategies for cell manipulation and skeletal tissue engineering using high-throughput polymer blend formulation and microarray techniques. Biomaterials. 2010;**31**(8):2216–28.

24. Thissen H, Johnson G, McFarland G *et al.*, eds. *Microarrays for the Evaluation of Cell–Biomaterial Surface Interactions*. 2007; Adelaide.

25. Goddard JM, Hotchkiss JH. Polymer surface modification for the attachment of bioactive compounds. Prog Polym Sci. 2007;**32**(7):698–725.

26. Bhat RR, Chaney BN, Rowley J, Liebmann-Vinson A, Genzer J. Tailoring cell adhesion using surface-grafted polymer gradient assemblies. Adv Mater. 2005;**17**(23):2802–+.

27. Rezaei SM, Ishak ZAM. The biocompatibility and hydrophilicity evaluation of collagen grafted poly(dimethylsiloxane) and poly (2-hydroxyethylmethacrylate) blends. Polym Test. 2011;**30**(1):69–75.

28. Zainuddin, Barnard Z, Keen I *et al.* PHEMA hydrogels modified through the grafting of phosphate groups by ATRP support the attachment and growth of human corneal epithelial cells. J Biomater Appl. 2008;**23**(2):147–68.

29. Chen H, Yuan L, Song W, Wu ZK, Li D. Biocompatible polymer materials: Role of protein-surface interactions. Prog Polym Sci. 2008;**33**(11):1059–87.

30. Saltzman WM, Kyriakides TR. Cell interactions with polymers. In: Lanza R, Langer R, Vacanti J, eds. *Principles of Tissue Engineering* 3rd edn: Elsevier; 2007. p. 279–96.

31. Matsumoto T, Mooney D. Cell instructive polymers. In: Lee K, Kaplan D, eds. *Tissue Engineering I*. Springer; 2006. p. 113–37.

32. Hook AL, Anderson DG, Langer R *et al.* High throughput methods applied in biomaterial development and discovery. Biomaterials. 2010;**31**(2):187–98.

33. Simon Jr CG, Sheng LG. Combinatorial and high-throughput screening of biomaterials. Adv Mater. 2011;**23**(3):369–87.

34. Kingshott P, Andersson G, McArthur SL, Griesser HJ. Surface modification and chemical surface analysis of biomaterials. Curr Opin Chem Biol. 2011;**15**, 667–76.

35. Dalby MJ, Gadegaard N, Tare R *et al.* The control of human mesenchymal cell differentiation using nanoscale symmetry and disorder. Nat Mater. 2007;**6**(12):997–1003.

36. Lin NJ, Lin-Gibson S. Osteoblast response to dimethacrylate composites varying in composition, conversion and roughness using a combinatorial approach. Biomaterials. 2009;**30**(27):4480–7.

37. Lin NJ, Drzal PL, Lin-Gibson S. Two-dimensional gradient platforms for rapid assessment of dental polymers: A chemical, mechanical and biological evaluation. Dental Mater. 2007;**23**(10):1211–20.

38. Thorstenson JB, Petersen LK, Narasimhan B. Combinatorial/high throughput methods for the determination of polyanhydride phase behavior. J Comb Chem. 2009;**11**(5):820–8.

39. Anderson DG, Tweedie CA, Hossain N *et al*. A combinatorial library of photocrosslinkable and degradable materials. Adv Mater. 2006;**18**(19):2614–18.

40. Urquhart AJ, Anderson DG, Taylor M *et al*. High throughput surface characterisation of a combinatorial material library. Adv Mater. 2007;**19**(18):2486.

41. Yang J, Mei Y, Hook AL *et al*. Polymer surface functionalities that control human embryoid body cell adhesion revealed by high-throughput surface characterization of combinatorial material microarrays. Biomaterials. 2010;**31**(34):8827–38.

42. Urquhart AJ, Taylor M, Anderson DG *et al*. TOF-SIMS analysis of a 576 micropatterned copolymer array to reveal surface moieties that control wettability. Analyt Chem. 2008;**80**(1):135–42.

43. Kurkuri MD, Driever C, Johnson G *et al*. Multifunctional polymer coatings for cell microarray applications. Biomacromolecules. 2009;**10**(5):1163–72.

44. Alexander MR, Taylor M, Urquhart AJ, Zelzer M, Davies MC. Picoliter water contact angle measurement on polymers. Langmuir. 2007;**23**(13):6875–8.

45. Kovalev A, Shulha H, Lemieux M, Myshkin N, Tsukruk VV. Nanomechanical probing of layered nanoscale polymer films with atomic force microscopy. J Mater Res. 2004;**19**(3):716–28.

46. Yune PS, Kilduff JE, Belfort G. Searching for novel membrane chemistries: Producing a large library from a single graft monomer at high-throughput. J Membrane Sci. 2012;**390**:1–11.

47. Zhou MY, Liu HW, Kilduff JE *et al*. High throughput membrane surface modification to control NOM fouling. Environ Sci Technol. 2009;**43**(10):3865–71.

48. Mei Y, Saha K, Bogatyrev SR *et al*. Combinatorial development of biomaterials for clonal growth of human pluripotent stem cells. Nat Mater. 2010;**9**(9):768–78.

49. Engler AJ, Sen S, Sweeney HL, Discher DE. Matrix elasticity directs stem cell lineage specification. Cell. 2006;**126**(4):677–89.

3 Materiomics using synthetic materials: metals, cements, covalent polymers and supramolecular systems

Björne B. Mollet, A. C. H. (Bram) Pape, Rosa P. Félix Lanao, Sander
C. G. Leeuwenburgh and Patricia Y. W. Dankers

Scope

To screen biomaterials in a materiomics approach, libraries of materials are produced. Different materials are used, varying from metals and cements, to covalent polymers that can be either premixed or polymerized *in situ*, to supramolecular systems that can be applied in a modular approach. This chapter describes the generation of such libraries using different kinds of materials and chemistries. Additionally, the advantages and limitations of the application of these different systems/biomaterials in a materiomics approach are discussed.

3.1 Introduction

Different synthetic biomaterials are used for many biomedical applications, varying from metals and ceramic cements, to polymers and supramolecular systems. To screen these biomaterials in a materiomics approach, as said above, libraries of materials are produced. Variations in biomaterials are screened as continuous gradients or in a discrete fashion. The properties that are varied and methods used to create variation within these libraries depend on the type of biomaterial. For the hard metal and ceramic-based biomaterials, the surface interaction with tissue is the property of most interest, and therefore properties such as surface roughness and topography are varied. Covalent polymers are diversified using combinatorial chemistry. The dynamic and self-assembling nature of supramolecular systems allows for the development of material libraries using a modular approach by mixing and matching of different compounds modified with supramolecular moieties.

Materiomics: High-Throughput Screening of Biomaterial Properties, ed. Jan de Boer and Clemens van Blitterswijk. Published by Cambridge University Press. © Cambridge University Press 2013.

Box 3.1 Classic experiment

Combinatorial screening of three-dimensional polymer–ceramic nanocomposites

Owing to the enormous complexity of the interaction between cells and metallic and bioceramic scaffold materials, there is a great need for rapid and systematic screening of these interactions in order to maximize the efficacy of tissue regeneration. These methods typically involve presentation of cells to biomaterials surfaces in a two-dimensional (2D) culture format, whereas cells *in vivo* reside in a three-dimensional (3D) extracellular matrix. Therefore, Chatterjee *et al.* recently developed combinatorial libraries of 3D porous scaffolds to screen the effect of the incorporation of calcium phosphate (CaP) nanoparticles in poly(ε-caprolactone) (PCL) scaffolds on cell response (1). They used combinatorial gradients and arrays were used to evaluate the effect of calcium phosphate content in 3D scaffolds on osteoblast adhesion and proliferation.

    Continuous PCL–CaP gradients were developed by combining two different polymer solutions containing either 100% PCL or 70% PCL, 30% CaP, using injection pumps. To generate a composition gradient, the flow rate of the PCL pump decreased linearly from 0.5 to 0 mL/min whereas the flow rate from the PCL–CaP pump increased to maintain a constant total effluent from the static mixer of 0.5 mL/min. In parallel, array libraries were created by collecting two drops of the effluent into wells of a 96-well flat-bottom polypropylene plate to study the effect of scaffolds with different PCL–CaP ratios in a more conventional manner. For both gradients and arrays, a salt leaching technique was used to generate porous scaffolds. By culturing MC3T3-E1 cells in contact with the gradients or arrays, the scaffold libraries were screened. Results revealed that higher calcium phosphate contents in 3D PCL–CaP porous scaffolds enhanced osteoblast differentiation, owing to release of calcium and phosphate ions which stimulate osteoblast function. Gradients were most suitable to study effects of scaffold composition on cells at short time points after cell seeding (e.g. cell adhesion) while arrays appeared to be more sensitive for measuring scaffold effects on cell proliferation at later time points. Summarizing, it was shown that combinatorial gradient approaches can be very useful for systematically studying the effect of scaffold composition on osteoblast response in 3D.

3.2   Basic principles of different synthetic materials

3.2.1   Metals and cements

Placing a metallic implant (such as titanium and its alloys) into the human body introduces a non-physiological surface into a physiological environment, resulting in a foreign body response that largely determines the final biological performance of such implants. This process is strongly controlled by the surface properties of the metallic implant in terms of (for example) surface roughness, topography, porosity and chemical composition. Metallic implant surfaces can be modified by altering the surface properties using physico-chemical treatments that affect topography and morphology. Examples of this type of surface modification include machining, polishing, grit-blasting, acid etching, alkali etching and anodization. Another strategy to modify the surface of metallic implants is to apply additional coatings to the implant surface to present biological cues to the surrounding tissue. In this way, for example, bone formation can be induced at the interface between the implant surface and the surrounding tissue.

Calcium phosphate cements (CPC) are commonly used in dentistry and orthopaedics as injectable bone void fillers. These cements have several advantages such as excellent biocompatibility due to their resemblance to the mineral phase in bone and teeth (70 wt% of native bone tissue consists of calcium phosphate nanocrystals). However, recent trends in bone tissue engineering aim to develop scaffolds that can be replaced in time by newly formed bone tissue, for example by inclusion of porogens that are able to degrade the calcium phosphate cement matrix by controlled production of acidity.

3.2.2 Synthetic covalent polymers

Polymers are macromolecules built from repeating units called monomers. The monomers are usually connected through covalent bonds that are formed during a polymerization reaction. The spread in molecular weight, which is expressed as the polydispersity index (PDI), is heavily influenced by the mechanism of polymerization. Synthetic polymers are in general more polydisperse than biopolymers such as proteins. Synthetic polymers can show a wide variation in their mechanical behaviour depending on the degree of crystallinity, degree of cross-linking and the values of the glass transition (T_g) and melting temperature (T_m) (see Chapter 2). The chain length influences the polymer strength, as longer polymers can form more entanglements between the chains. The incorporated monomer, of which a wide variety is commercially available, also has a large influence. Rigid, linear polymer backbones are able to pack together more closely in ordered, crystalline regions than branched chains. Polymers with higher crystallinity are stronger and less flexible than completely amorphous polymers. In the glassy state, below T_g, the entangled chains in the amorphous regions are 'frozen' and unable to move, resulting in a more brittle polymer. Above the T_g, the chains in the amorphous regions gain the freedom to move, making the polymer more flexible. At T_m the crystalline regions melt and the polymer becomes a viscous liquid. For polymer biomaterials to be used inside the human body, a T_g or T_m above or below 37 °C are thus important design criteria. The introduction of a plasticizer can lower both the T_g by allowing the chains to slide over each other more easily, and the T_m by preventing crystallization. Cross-linking of polymer chains via covalent bonds makes a polymer harder and more difficult to melt.

It should be clear that polymeric materials display an extraordinary range of variables that can be used to diversify the material and tune the physical properties.

3.2.3   Supramolecular chemistry and (bio)materials

Supramolecular chemistry can be described as chemistry beyond the molecule and is based on non-covalent interactions between molecules (2). These interactions include metal coordination, hydrophobic forces, hydrogen bonding, pi–pi interactions, electrostatic effects and van der Waals interactions. Non-covalent interactions are weaker than covalent bonds. Owing to the lower bond energies of non-covalent interactions, supramolecular chemistry gives rise to intrinsically dynamic materials. In these materials, not only spatial (structural) but also temporal (dynamical) features play an important role in the material properties (3). Inspired by the synthetic pathways and molecular self-assembly mechanisms by which natural materials are produced, supramolecular chemistry has opened new perspectives in (bio)material science. The accessibility of self-assembled synthetic biomaterials has introduced a rich variety of novel architectures and material properties.

Self-assembly is based on information stored at the molecular level. Molecules can be 'programmed' for molecular recognition via specific spatial relationships which give rise to complementary geometries. When combined with non-covalent interaction patterns such as hydrogen bonding arrays, sequences of donor and acceptor groups, and metal

---

Box 3.2 Classic experiment

Combinatorial polymer chemistry screening

Large libraries of different covalent polymers can be generated using combinatorial chemistry to quickly screen for a potential useful biomaterial. Anderson *et al* (4). used 25 monomers to prepare 576 discrete mixtures in triplicate in a microarray format (Figure 3.1A). Cells were seeded on top of the array to study the effect on stem cell differentiation.

All monomers were dissolved in dimethylformaldehyde (DMF), and for all possible combinations premixed in a ratio of 70:30 vol% in a 384-well plate. Sequentially, the mixtures were deposited on a poly (z-hydroxyethyl methacrylate) (pHEMA) coated glass slide. The pHEMA inhibits cell growth but provides a layer on which the monomer mixtures can be polymerized. To prevent spreading of the droplets, the mixtures were exposed to ultraviolet light to polymerize the monomers after each deposition round. In total, 1728 spots of monomer mixtures were printed on a single microscope slide and polymerized by UV. Several microscope slides were stained and imaged simultaneously. A drawback of the method is that owing to washing and differences in monomer reactivity the exact composition of the polymer in each spot is unknown. Next, cells were deposited on top of the slides and their ability to adhere was tested. Because standard microscope glass slides were used, cell visualization was easily performed using a regular fluorescence microscope (Figure 3.1B). In this study, the quantitative analysis was done by hand. However, high-content image analysis could be adapted for the analysis of these microarrays.

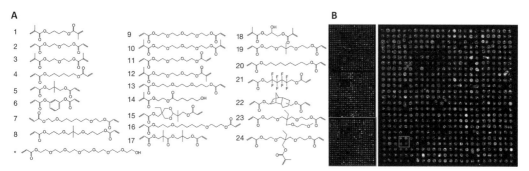

Figure 3.1   A, The 25 different monomers used for the preparation of 576 chemically different polymers. B, On a single microscopy slide 3 × 576 polymer spots are deposited. Cells cultured on the polymer microarray are visualized with fluorescence microscopy (4).

coordination groups, self-assembly can be induced by design (3). In addition, the collective of multiple weak non-covalent interactions can give rise to more stable, yet still dynamic self-assembled structures. The dynamic nature gives supramolecular materials the ability to associate and dissociate reversibly, and be reconstructed and deconstructed, enabling these materials to be responsive to environmental factors such as temperature and pH and to interact with external entities such as cells (3). These properties make supramolecular materials eminently suitable to create multifunctional biomaterials for tissue engineering that more closely mimic the natural counterparts that they aim to substitute or interact with. Furthermore, the dynamic rearrangement of the molecular components allows for the incorporation of (bio)active molecules. This has proved useful for bioactivation of materials via induced incorporation of, for example, extracellular matrix (ECM)-derived peptide sequences (5, 6), or for the establishment of controlled drug release profiles in the field of drug delivery devices (7).

3.3   Materiomics using synthetic materials

3.3.1   Screening of metals and cements

A library of both organic (such as peptides and proteins) and inorganic coatings (such as calcium phosphate) to cover titanium implants has been developed over the past decade (8, 9). These surface treatments are characterized by a wide variety of physico-chemical characteristics such as roughness, hydrophilicity and (nano)texture, while coatings can be produced with a virtually unlimited range of parameters with respect to thickness, crystal phase and degree of crystallinity, elemental composition, presence of dopants etc. All of these parameters will affect biological phenomena such as protein adsorption, cell attachment and thus the final biological behaviour *in vivo*.

In the field of dental and bone cements, various studies have focused on the evaluation of different CPC formulations in order to achieve progressive degradation and replacement by the host tissue. For example, arrays have been studied to identify the optimal nature of a porogen mixed into the CPC to induce porosity (sugar crystals, salt crystals, foaming agents, natural or synthetic polymers) (10–12). Porogens can also be used to degrade the CPC matrix by controlled production of acidity. Studies have aimed to reveal the ideal degradation rate of a CPC porogen in order to allow new bone formation without compromising the 3D structure while creating sufficient space for tissue ingrowth. Poly(lactic-co-glycolic) acid (PLGA) microspheres have been widely studied as CPC porogens, and several CPC–PLGA libraries have been developed with tailorable degradation properties which can be applied for different clinical needs (13).

Box 3.3 Classic experiment

Blending of polymers and gradient screening

Screening a range of different mixtures is also possible using a materiomics approach. Using a continuous gradient from 100% polymer A to 100% polymer B it is possible to identify the ideal mixing ratio very fast. This has been done by Simon *et al.* using blends of poly(L-lactic acid) (PLLA) and poly(D,L-lactic acid) (PDLLA) (14). Both polymers are commercially available, and were dissolved in a common solvent before being loaded into two syringes. Next, the polymers were mixed while simultaneously a third syringe withdrew small volumes to obtain a composition gradient inside this needle (Figure 3.2). The content of this syringe was deposited on the substrate, from which a 4 µm thick film was then prepared using a motorized knife. As a control, five different discrete blends were prepared. All films were melted above the polymer melting temperatures and annealed at 120 °C for 8 h to remove the solvent and induce crystallization. This is below the melting temperature but above the glass transition temperature, allowing some flexibility in the polymers. Infrared spectroscopy and atomic force microscopy (AFM) were used to investigate and map the gradients. Cells were cultured on the surfaces and cell numbers were automatically counted. Adhesion of the cells was similar over the whole range, but proliferation was faster on the PDLLA-rich ends of the gradient.

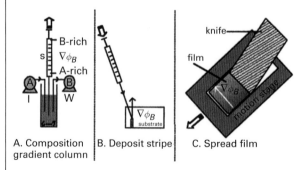

A. Composition gradient column | B. Deposit stripe | C. Spread film

Figure 3.2 The three stages in preparing a gradient film of two polymers (15).

3.3.2   Arrays of synthetic covalent polymers

Combinatorial chemistry has opened ways to create and explore new polymers that comply with the needs of specific biomedical applications. Besides the fast screening, combinatorial methods have further advantages including faster data acquisition, more thorough examination of experimental variables, equal processing conditions for a given specimen and lower experimental error. Combinatorial chemistry was developed by the pharmaceutical industry and has led to great advances in the search for new drugs. However, the synthesis and screening techniques used for pharmaceuticals could not be directly applied in polymer chemistry. One key point is that in drug discovery it is possible to identify useful lead compounds contained within a complex mixture. In polymer discovery, it is impossible to screen for useful material properties unless the test polymer can be obtained in a state of high purity (16). Scientists have adapted these techniques to be able to use combinatorial chemistry for the study of polymer material properties.

Here, we discuss three different methods of varying polymer properties in a combinatorial approach, each illustrated by examples from literature. The first method involves the effect of variation in monomer composition. A second method to change the properties is the mixing of two different polymers. In the third method, besides the chemical composition of the material, the processing of polymers also influences the final bulk properties. Variations in biomaterials are screened as gradients or in a discrete fashion. Additionally, both 2D (surface) and 3D (bulk) morphologies are of interest.

Variations in monomer composition

Because a polymer consists of many monomers, it is possible to synthesize polymers consisting of multiple monomers with varying properties, such as different side chains (as indicated with R_2 in Figure 3.3). These side chains can have a profound effect on the chemical and material properties of the polymer. Furthermore, in such a copolymerization the organization of the different monomers can be varied. For example in a copolymer consisting of monomer A and B, the monomers can alternate $(AB)_n$, form blocks (A_nB_m), or have a periodic (e.g. $(AABABBB)_n$) or randomized arrangement.

Polyacrylates

Polyacrylates are prepared via photo-initiated polymerization, which is a popular method for the preparation of biomaterials because of the controlled reaction, short reaction time and gentle conditions (Figure 3.3A) (17). This allows for the introduction of biological compounds, and reactions can be done *in vivo*. Furthermore, a variety of monomers is commercially available. Anderson *et al.* (4) used 25 acrylate-monomers selected on commercial availability to prepare 576 different polymers in a microarray format. This experiment is explained in more detail in Box 3.2.

Takeuchi *et al.* (18) used a molecular imprinting technique to quickly scan for an artificial receptor. In this technique, the monomers are polymerized in the presence of a template molecule. After removal of the template molecule, the polymer network had

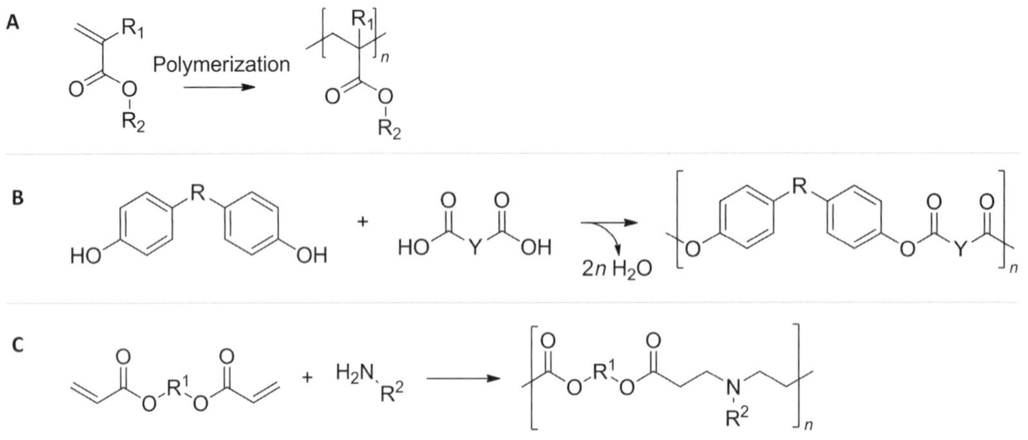

A

B

C

Figure 3.3   A, General structure of the acrylate monomer and the polyacrylate. B, Condensation of diphenol with diacid leads to the polyarylate. The synthesis is often performed via acid chloride. C, General synthesis of poly(β-amino ester)s via the conjugate addition of an amine to a diacrylate.

formed a cavity that could function as an artificial receptor for the templated molecule. Using a programmed liquid handler, glass vials were filled with a pre-polymer mixture of the acrylate monomers, initiators and templates. The monomers, methacrylic acid and the more acidic 2-(trifluoromethyl)acrylic acid, were mixed in different ratios together with (ethylene glycol) dimethacrylate as cross-linker. The mixture was polymerized by exposure to UV light in the presence of a template molecule, which was removed afterwards.

The material properties of dental restorative composites are influenced by a large number of parameters, and therefore a high-throughput combinatorial approach is ideal. (19) Two-dimensional samples were prepared of which one dimension is continuous and one dimension is discrete (Figure 3.4A). For the discrete gradient, ethoxylated bisphenol-A dimethacrylate and tri(ethylene glycol) dimethacrylate (EGDM) were mixed in different ratios to study the effect of the content of bisphenol-A present in the material. By using different UV irradiation times, a methacrylate conversion gradient was created along the other dimension while studying the required minimal degree of polymerization. A similar approach was taken for bisphenol-A dimethacrylate (not ethoxylated) Figure 3.4B (20).

The objective of Pieper *et al.* (21) was to relate the bulk properties of siloxane-polyurethane to the polymer's performance as anti-fouling coating by varying the acrylic polyols. The coating was prepared by reacting 3-aminopropyl-terminated poly(dimethylsiloxane) (PDMS) with the polyols and a polyisocyanate cross-linker. Butyl methacrylate (BMA), n-butyl acrylate (BA) and 2-hydroxyethyl acrylate (HEA) were mixed in different ratios. HEA was used to vary the cross-linking density, and the ratio of BMA to BA was used to tune the T_g of the polymer. Using a dispensing robot, 24 reactions were performed simultaneously. The monomer mixture was polymerized at 95 °C for 10 hours and the mixture could be used without further purification. The effect of the PDMS and presence of poly(ε-caprolactone) (PCL) in another layer of this coating was also studied in a high-throughput fashion (22).

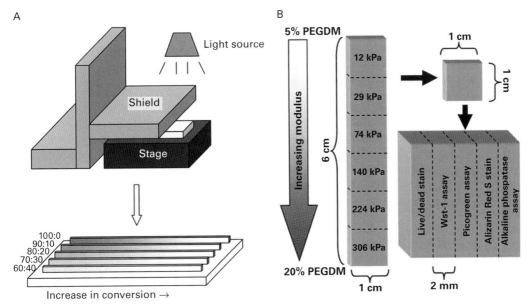

Figure 3.4   Examples of gradient preparations. A, Preparation of a 2D sample with one discrete and one continuous gradient.(19) B, A gradient sample is divided to perform several different stainings and assays.(20) PEGDM: poly(ethylene glycol) dimethacrylate.

Hydrogels based on polyacrylates have also been studied, using microfluidics in combination with photopolymerization (23). Varying concentrations of poly(ethylene glycol) diacrylates (PEGDA) with different molecular weights (1 kDa and 4 kDa), were used to yield different cross-link densities, and 1 kDa PEGDA was mixed with an adhesive acrylated protein to yield hydrogels with different levels of cell attachment. Different concentrations of pre-polymer solutions of PEGDA have also been mixed with cells, transferred to a mould and photopolymerized (24). The stiffness of the material increased with increasing PEG content (Figure 3.4B). Staining with different agents showed that variation of the modulus of the hydrogel is enough to induce changes in cell responses.

Tyrosine-derived polymers

Pseudo-poly(amino acids) are promising biomaterials because of their favourable properties such as biodegradation into non-toxic and easily metabolizable products. A series of four polycarbonates derived from the ethyl, butyl, hexyl and octyl esters of desaminotyrosyl-tyrosine was prepared by condensation polymerization (Figure 3.3B) (25). The increase in chain length of two carbons per step allowed for the tunability of the T_g, solubility and surface properties such as hydrophobicity and cell adhesion.

Brocchini *et al.* (26) used 14 different tyrosine-derived diphenols and eight aliphatic esters to synthesize a library of 112 polyarylates. Such a library could be used on the one hand to identify new hits for specific applications, and on the other hand for the structural investigation of structure–property relationships. They found that variations in the structure led to changes in free volume, flexibility and hydrophobicity. Owing to the similar reactivity of the monomers, 32 reactions could be performed simultaneously in a water bath. For all samples the T_g and the surface-contact angle were determined and cell growth was studied.

Polyesters

Poly(β-amino esters) show low cytotoxicity and are easily synthesized by the conjugate addition of a primary amine or bis(secondary amine) to a diacrylate (Figure 3.3C). Therefore, synthesis is easily done, and 2350 polymers were synthesized within a day and tested at a rate of 1000 per day (21). Based on these results, a library of over 500 degradable poly(β-amino esters) was used to find optimal transfection potential and biocompatibility (27). This work was combined with the photopolymerizable acrylate monomers to combine biocompatibility with good mechanical properties (28). First, a library of 120 macromers based on 12 amines and 10 diacrylates was prepared. Subsequently, 89 of these macromers were polymerized, resulting in networks that displayed a wide range of degradation times and mechanical properties.

Polyurethanes

Another class of polymers that has been investigated is the class of the polyurethanes (29). Polyurethanes are used in many applications, such as medical devices. Polymerizing polyol, diisocyanate and chain extenders led to a library of 280 polyurethanes which have been chemically characterized. A big influence of the concentration and hydrophilicity of the polyol was observed.

Polymer blends

Another method of optimizing material properties, such as the T_g, mechanical properties and biocompatibility, is to blend polymers. This is a common and inexpensive method which is used for many manufactured polymers. A popular biopolymer used in blends at the National Institute of Standards and Technology (NIST) is poly(lactic acid), which can be derived from biomass. The polymer can be made in different stereochemistries, which gives the ability to tune T_m and T_g. First, a technique was developed to generate 2D gradient polymer blends (15). Polystyrene and poly(vinyl methyl ether) were mixed in toluene and placed as a stripe gradient on silicon. A knife was used to spread the stripe in the orthogonal direction, and a temperature gradient was applied for annealing of the composite. Cloud point and phase separation could then be studied.

   In later work, this gradient-library method was applied to study the surface properties of biodegradable polymer blends (30). Poly(D,L-lactic acid) (PDLLA) and PCL were mixed, both polymers being approved in certain devices by the Food and Drug Administration (FDA). A surprising combination of mixture composition and annealing temperature was found that had a profound effect on alkaline phosphatase expression. The stereochemistry of PLA was further exploited in mixing poly(L-lactic acid) (PLLA) and PDLLA (14). Although both polymers are chemically similar, they have a different tacticity, and as a result PLLA is crystalline and has a higher modulus whereas PDLLA is amorphous. This experiment is explained in more detail in Box 3.3.

Besides these polymer films, 3D scaffold gradients have been prepared using PDLLA (31). As a proof-of-principle, a mixture of Sudan IV ('red') loaded PDLLA and clear PDLLA solutions were loaded into a mould containing NaCl crystals. After lyophilization of the scaffolds, the NaCl porogen was leached out with water to create a porous 3D scaffold. In further research, different polymers were used to test the effect on cell function.

The tyrosine-derived polymers used for varying monomer compositions have been used in blends. Films made of blended poly(desaminotyrosyl-tyrosine ethyl ester carbonate) (pDTEc) and poly(desaminotyrosyl-tyrosine octyl ester carbonate) (pDTOc) were subjected to a gradient temperature protocol (32). Using the above described NaCl-leaching method in 96-well plates, pDTEc and pDTOc have also been used to create 3D scaffolds (33). Although the two polymers only differ in the length of the side chain, they have different hydrophilicity, T_g, mechanical properties and degradation rates. These properties affect cellular responses and gene expression. High-throughput screening allowed for the rapid identification of the most effective combinations.

Processing of gradients

Besides differences in interactions caused by the chemical composition of the materials, physical properties are also determined by the way the polymer is processed, as we already have seen in some examples where annealing temperature gradients were used. The interaction of cells with material surfaces is often induced by ECM components adsorbed on the surface of the materials. Therefore, protein surface adsorption and resulting cell responses have been studied using high-throughput methods.

Bhat *et al.* (34) used pHEMA because of the anti-fouling properties and the ease of synthesis via radical polymerization. It is less efficient in protein repellency than poly-ethylene oxide (PEO), and therefore offers a broad range in which protein adsorption can be modified. A gradient in molecular weight was prepared by slowly pulling a wafer out of the monomer solution, thereby creating differences in residence times in solution. An orthogonally grafted density gradient was created by an initiator gradient on the wafer perpendicular to the pulling direction. Fibronectin (FN) was deposited on the surface, and the thickness of FN gradually changed in the opposite direction to the increasing pHEMA thickness. The difference in FN surface coverage also led to a difference in cell adhesion and spreading. Furthermore, a materiomics approach where many conditions can be tested simultaneously can help in modelling and predicting the adsorption of FN to the surface (35).

Poly(2-hydroxyethyl methacrylate) is not the only substrate that has been used to study the adsorption of proteins. Ionov *et al.* (36) prepared a layer of poly(glycidyl methacrylate) (PGMA) and annealed PEG on top of this layer using a temperature gradient, to create a 1D variation in grafting density. Next, kinesin motor proteins were deposited on the PEG, followed by microtubules which showed mobility and size-sorting on this gradient surface. Phospholipid-containing polymers have also been used to study protein adsorption and fibroblast cell adhesion (37). Monomers of ω-methacryloyloxyalkyl phosphorylcholine (MAPC) were polymerized onto a polyethylene sheet which contained a gradient of peroxide moieties prepared by treatment with a radiofrequency corona-discharge apparatus. Increasing the length of the alkyl spacers allowed the tuning of the hydrophobicity and thereby the material properties.

3.3.3   Screening of supramolecular biomaterials

Several synthetic supramolecular (bio)materials have been developed over the past decades, varying from amphiphilic block copolymers to peptide amphiphiles and true supramolecular polymers. However, none of these systems have been used in a materiomics approach. Therefore, here, three different systems will first be discussed, after which a perspective on the applicability of supramolecular chemistry in high-throughput screening methods will be given.

Examples of supramolecular systems

In solution, block copolymers consisting of a hydrophilic (polar) part A and a hydrophobic (apolar) part B will self-assemble into micelles owing to their amphiphilic nature. This property makes amphiphilic materials suitable as a drug delivery vehicle of, for instance, water-insoluble drugs (38). Amphiphilic molecules are also known to give rise to other structures such as vesicles and 3D networks (7). These networks can contain high amounts of water, forming so-called hydrogels which have potential for mimicking the natural ECM. Other low-molecular-weight compounds that form hydrogels based on self-assembly can be rationally designed (39).

Peptides have also been used to form amphiphilic structures. These peptide amphiphiles (PA) have been applied to form injectable nanofibrous hydrogels. Peptides are of particular interest in the formation of new biomaterials because of the large variety of short sequences that can easily be made by automated chemical synthesis, and because of their potential for bioactivity. For instance, Silva *et al.* developed an artificial nanofibrous PA scaffold that presented a very high density of neurite-promoting laminin epitope IKVAV, which induced very rapid differentiation of cells into neurons relative to laminin or soluble peptides (Figure 3.5A, B) (6).

As previously discussed in this chapter, covalent polymers form an important group of biomaterials. Their high molecular weight gives rise to advantageous mechanical properties. Analogous, supramolecular polymerization can occur, but the chain length that can be formed is dependent on the average lifetime of the reversible bonds. The self-complementary unit 2-ureido-4[1H]-pyrimidinone (UPy) dimerizes via four hydrogen bonds (Figure 3.5C), which together give rise to a relative long lifetime of 0.1–1 seconds. The work of Sijbesma *et al.* demonstrated how monomers containing two of these units self-assemble by chain extension into linear supramolecular polymers with a virtual high molecular weight (Figure 3.5D), whereas three or more of these units in a monomer results in polymer networks (40). Depending on the backbone of the supramolecular

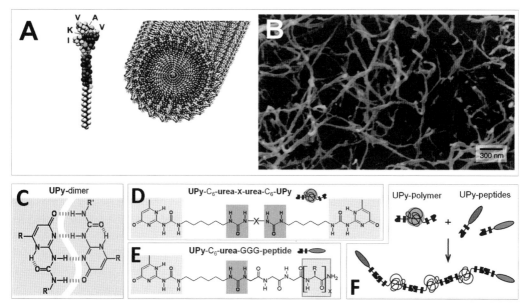

Figure 3.5 A, Molecular graphics illustration of an IKVAV-containing peptide amphiphile molecule and its self-assembly into nanofibres. B, Scanning electron micrograph of an IKVAV nanofibre network formed by adding cell media to a peptide amphiphile aqueous solution (6). C, The self-complementary 2-ureido-4[1H]-pyrimidinone (UPy) dimerizes via four intermolecular hydrogen bonds. D, The chemical structure of a polymer (X) which is end-functionalized with the UPy-moiety (UPy-polymer). E, The chemical structure of a peptide functionalized with a UPy-moiety (UPy-peptide). F, UPy-polymer mixed with UPy-peptides results in a bioactive supramolecular polymer.

monomer the polymerization results in a hydrogel, for instance when poly(ethylene glycol) is used (41), or a 'hard' material which is the case for polycaprolactone (Figure 3.5D, i.e. X in structure) (42). Bioactivity in these materials was achieved via cell binding peptides that were incorporated in the material using the same quadruple hydrogen bonding unit (Figure 3.5E, F) (5, 43). The material properties can be easily changed by mixing different supramolecular polymers, resulting in a variety of mechanical properties (44).

Perspective on supramolecular chemistry in materiomics
Supramolecular chemistry holds both opportunities and challenges in relation to materiomics. Its dynamic nature introduces a new perspective on material diversity and the generation of libraries. Supramolecular systems are eminently suitable for combinatorial approaches as supramolecular synthesis allows for a mix-and-match principle, i.e. a modular approach, as long as the requirements for self-assembly are met. Nevertheless, to this date no examples of supramolecular biomaterial library screening can be found in literature. An example of screening using supramolecular interactions which is not

biomaterial-related is the identification of ligands using the non-covalent, dynamic nature of the interactions between biological molecules. In phage display and systematic evolution of ligands by exponential enrichment (SELEX) (45), the selection of the best ligand candidates from large protein libraries is based on the strongest non-covalent interactions.

The characterization of materials with a dynamic and responsive nature is not straight-forward. It demands deeper and more deliberate analysis, taking thermodynamic versus kinetic aspects of the materials into account. For these biomaterials it is even more important than for covalent materials to perform analysis under conditions that are relevant to the final application. For example, body temperature when compared with room temperature can have a large effect on the outcome of the material properties that are measured. One should even consider possible effects of the characterization method itself on the material. For instance, for a material that is built up from molecules connected through hydrogen bonds, water contact angle measurements can cause reorganization of molecules at the surface, and thus the characterization method itself influences the material property that is measured. Both standardization and characterization methods will have to be revised before a materiomics approach can unleash the full potential of this new generation of biomaterials.

3.4   Future perspectives

The properties of ceramics, metals and materials made out of synthetic polymers or supramolecular systems can be varied in multiple different ways as we have shown. Covalent materials show the potential of modification by combinatorial chemistry. The polymer can be chemically varied by choosing monomer and monomer ratios before polymerization. Other options to change material properties are blending different polymers and altering the processing of the materials. Materiomics allows for the rapid screening and the careful investigation of parameter spaces of the many parameters influencing the success of a new biomaterial.

Another way to study combinatorial material libraries is the use of computational models. The complicated interactions between biomaterials and living cells or tissues are difficult to capture in appropriate computational models, but continuous progress in computational modelling techniques has led to the use of computational methodologies in the field of biomaterials discovery. Virtual combinatorial libraries have shown to be a time- and cost-effective means to explore a wide range of new material compositions. The properties of polymers within a virtual library are predicted by molecular modelling tools. Based on a list of desired properties, this allows the rational selection of a smaller subset of promising polymer compositions that can be further explored via actual synthesis. An example in which computational modelling techniques have aided the discovery of a new biomaterial through rational design is the first degradable radio-opaque polymer for coronary stents. In this discovery process the material requirements were defined first, followed by a targeted search within a virtual library of approximately 10 000 distinct polymer compositions, using semi-empirical modelling techniques. Only a small subset of candidate materials needed to be synthesized and experimentally explored. It took less than 9 months to identify the polymer with optimal material properties for the specific application (46).

Furthermore, it has been proposed that supramolecular biomaterials will be eminently suitable as, for example, synthetic extracellular matrices, because of their dynamic nature which mimics the situation *in vivo*. By using a modular approach it is proposed that different components can be mixed and matched, resulting in large arrays of supramolecular biomaterials with different material properties, e.g. differences in bioactivity, mechanical behaviour or non-fouling properties. The dynamic nature of these systems is important, as it means they will change in time or as a reaction to changes in the environment, including during analysis. However, this also occurs in natural systems which are intrinsically dynamic. Additionally, covalent systems may also change, a point that is often neglected in biomaterials research using covalent synthetic and/or natural polymers. For example, in natural protein-based biomaterials such as collagen or silk the secondary protein structure plays a key role in the specific material properties, but this structure is based on non-covalent interactions. Other examples are synthetic polymeric materials, whose monomeric components are linked through reversible covalent bonds (47, 48).

3.5   Snapshot summary

- For the hard metal and ceramic-based biomaterials, primarily the surface interaction with tissue is of interest. The diversity within the material libraries that are studied therefore originates from variations in surface roughness, topography, porosity and chemical composition of coatings.
- Within polymer chemistry there are many variables that can be used to diversify material properties. Examples are monomer choice/design, side groups, branching, cross-linking, copolymerization and polymer mixing. Other important parameters are the polymer molecular weight and molecular weight distribution.
- Besides polymer chemistry, polymer processing affects the properties of a polymer and thus creates another parameter set which can be used to optimize material properties. Because not all effects of the processing on polymers are known, screening provides a valuable tool.
- Supramolecular biomaterials are based on non-covalent interactions between molecules. These materials are therefore intrinsically dynamic; not only spatial (structural) but also temporal (dynamical) features play an important role in the material properties.
- The possibilities that supramolecular chemistry offers to create new (libraries of) biomaterials is immense, but the exploration of their true potential will rely on the development of suitable (high-throughput) characterization and standardization methods that respect the dynamic aspect of these materials.

- Dynamics as obviously present in supramolecular (bio)materials are often neglected in biomaterials research using covalent synthetic and/or natural polymers. It is important to acknowledge that these covalent systems may also change in time.
- Materials can generally be processed into several physical forms, such as bulk material, coatings, films, micro/nanoparticles, gels, micro/nanofibres or membranes. The relevant form for the final application should be considered when performing a material screen.
- The question of gradients versus discrete screening needs to be considered. Discrete screening allows for the fast identification of interesting candidates from large material libraries, whereas gradients enable the identification of the optimal match for a certain application. This is relevant as small differences in material properties have proved to have a big impact on the behaviour of cells.
- Gradients are relevant not only for screening of biomaterials, but also in the end-product when used in tissue engineering, as many tissues display strong non-homogeneous characteristics.

Further reading

Collier JH, Rudra JS, Gasiorowski JZ, Jung JP. Multi-component extracellular matrices based on peptide self-assembly. Chem Soc Rev. 2010;**39**(9):3413–24. This tutorial review describes the approaches that have been taken to make synthetic extracellular matrices. A diversity of strategies that are based on molecular self-assembly is compared with the structures and processes in native ECM.

Corbett PT, Leclaire J, Vial L, West KR *et al.* Dynamic combinatorial chemistry. Chem Rev. 2006;**106**(9):3652–711 presents a comprehensive review of all aspects of dynamic combinatorial chemistry and some related approaches. It includes a detailed discussion of the various types of reversible chemistries, experimental setups and applications, and is not limited to biomaterials.

Dankers PYW, Meijer EW. Supramolecular biomaterials. a modular approach towards tissue engineering. Bull Chem Soc Japan. 2007;**80**(11):2047–73 illustrates the strength of the modular approach to easily diversify material properties of supramolecular polymers. Via a simple 'mix-and-match' principle, mechanical properties are tuned (see also refs (42) and (44)).

Fisher JP, Dean D, Engel PS, Mikos AG. Photoinitiated polymerization of biomaterials. Ann Rev Mater Res. 2001;**31**(1):171–81 reviews the photopolymerization for polymers typically suited for biomaterials. Furthermore, studies on these materials related to biomaterials and cell interactions are discussed.

Le Guéhennec L, Soueidan A, Layrolle P, Amouriq Y. Surface treatments of titanium dental implants for rapid osseointegration. Dental Mater. 2007;**23**(7):844–54 shows an overview and discusses future trends of the different methods used for increasing surface roughness or applying coatings to improve osseointegration of titanium dental implants.

Lee J-H, Gu Y, Wang H, Lee WY. Microfluidic 3D bone tissue model for high-throughput evaluation of wound-healing and infection-preventing biomaterials. Biomaterials. 2012 Feb;**33**(4):999–1006. This paper describes the development of a microfluidic 3D bone tissue model to assess the efficacy of biomaterials aimed at accelerating implant wound healing and preventing bacterial infection.

Kohn J, Welsh WJ, Knight D. A new approach to the rational discovery of polymeric biomaterials. Biomaterials. 2007;**28**(29):4171–7. This is a leading opinion paper that illustrates both the need for new approaches to biomaterials discovery as well as the significant promise inherent in the use of combinatorial synthesis, high-throughput experimentation, and computational modelling.

Moulin E, Cormos G, Giuseppone N. Dynamic combinatorial chemistry as a tool for the design of functional materials and devices. Chem Soc Rev. 2012;**41**(3):1031. This tutorial review illustrates the possibilities that supramolecular and dynamic covalent systems provide to develop new materials and devices via the application of dynamic combinatorial chemistry.

Singh M, Berkland C, Detamore MS. Strategies and applications for incorporating physical and chemical signal gradients in tissue engineering. Tissue Eng Part B: Rev. 2008;**14**(4):341–66. This review presents an overview of key methodologies to generate all kinds of gradients in a diversity of biomaterials and their possible implication for tissue engineering applications.

Viney C. Processing and microstructural control: lessons from natural materials. Mater Sci Rep 1993;**10**(5):187–236. This review demonstrates how material science can benefit from detailed knowledge of the synthetic pathways and molecular self-assembly mechanisms by which natural materials are produced, by describing the most significant classes of natural macromolecules.

References

1. Chatterjee K, Sun L, Chow LC, Young MF, Simon CG Jr. Combinatorial screening of osteoblast response to 3D calcium phosphate/poly(ε-caprolactone) scaffolds using gradients and arrays. Biomaterials. 2011;**32**(5):1361–9.
2. Lehn J-M. *Supramolecular Chemistry: Concepts and Perspectives*. VCH; 1995.
3. Lehn J-M. From supramolecular chemistry towards constitutional dynamic chemistry and adaptive chemistry. Chem Soc Rev. 2007;**36**(2):151–60.
4. Anderson DG, Levenberg S, Langer R. Nanoliter-scale synthesis of arrayed biomaterials and application to human embryonic stem cells. Nat Biotechnol. 2004;**22**(7):863–6.
5. Dankers PYW, Boomker JM, Huizinga-van der Vlag A *et al.* Bioengineering of living renal membranes consisting of hierarchical, bioactive supramolecular meshes and human tubular cells. Biomaterials. 2011;**32**(3):723–33.
6. Silva GA, Czeisler C, Niece KL *et al.* Selective differentiation of neural progenitor cells by high-epitope density nanofibers. Science. 2004 27;**303**(5662):1352–5
7. Branco MC, Schneider JP. Self-assembling materials for therapeutic delivery. Acta Biomater. 2009;**5**(3):817–31.
8. Jonge LT, Leeuwenburgh SCG, Wolke JGC, Jansen JA. Organic–inorganic surface modifications for titanium implant surfaces. Pharmaceut Res. 2008 **29**;25(10):2357–69.
9. Le Guéhennec L, Soueidan A, Layrolle P, Amouriq Y. Surface treatments of titanium dental implants for rapid osseointegration. Dental Mater. 2007;**23**(7):844–54.
10. Xu HHK, Takagi S, Quinn JB, Chow LC. Fast-setting calcium phosphate scaffolds with tailored macropore formation rates for bone regeneration. J Biomed Mater Res Part A. 2004;**68**A(4):725–34.
11. Liao H, Walboomers XF, Habraken WJEM *et al.* Injectable calcium phosphate cement with PLGA, gelatin and PTMC microspheres in a rabbit femoral defect. Acta Biomater. 2010;**7**(4):1752–9.

12. Perut F, Montufar EB, Ciapetti G *et al*. Novel soybean/gelatine-based bioactive and injectable hydroxyapatite foam: Material properties and cell response. Acta Biomater. 2011;**7**(4):1780–7.

13. Félix Lanao RP, Leeuwenburgh SCG, Wolke JGC, Jansen JA. In vitro degradation rate of apatitic calcium phosphate cement with incorporated PLGA microspheres. Acta Biomater. 2011;**7**(9):3459–68.

14. Simon J, Eidelman N, Kennedy SB, Sehgal A, Khatri CA, Washburn NR. Combinatorial screening of cell proliferation on poly(L-lactic acid)/poly(D, L-lactic acid) blends. Biomaterials. 2005;**26**(34):6906–15.

15. Meredith JC, Karim A, Amis EJ. High-throughput measurement of polymer blend phase behavior. Macromolecules. 2000;**15**;33(16):5760–2.

16. Kohn J, Welsh WJ, Knight D. A new approach to the rational discovery of polymeric biomaterials. Biomaterials. 2007;**28**(29):4171–7.

17. Fisher JP, Dean D, Engel PS, Mikos AG. Photoinitiated polymerization of biomaterials. Ann Rev Mater Res. 2001;**31**(1):171–81.

18. Takeuchi T, Fukuma D, Matsui J. Combinatorial molecular imprinting: an approach to synthetic polymer receptors. Analyt Chem. 1998 Dec 11;**71**(2):285–90.

19. Lin-Gibson S, Landis FA, Drzal PL. Combinatorial investigation of the structure–properties characterization of photopolymerized dimethacrylate networks. Biomaterials. 2006 Mar; **27**(9):1711–7.

20. Anderson DG, Lynn DM, Langer R. Semi-automated synthesis and screening of a large library of degradable cationic polymers for gene delivery. Angew Chem Int Ed. 2003;**42**(27):3153–8.

21. Pieper R, Ekin A, Webster D *et al*. Combinatorial approach to study the effect of acrylic polyol composition on the properties of crosslinked siloxane-polyurethane fouling-release coatings. J Coatings Technol Res. 2007 Dec 1;**4**(4):453–61.

22. Ekin A, Webster D, Daniels J *et al*. Synthesis, formulation, and characterization of siloxane-polyurethane coatings for underwater marine applications using combinatorial high-throughput experimentation. J Coatings Technol Res. 2007 Dec 1;**4**(4):435–51.

23. Burdick JA, Khademhosseini A, Langer R. Fabrication of gradient hydrogels using a micro-fluidics/photopolymerization process. Langmuir. 2004;**28**;20(13):5153–6.

24. Chatterjee K, Lin-Gibson S, Wallace WE *et al*. The effect of 3D hydrogel scaffold modulus on osteoblast differentiation and mineralization revealed by combinatorial screening. Biomaterials. 2010;**31**(19):5051–62.

25. Ertel SI, Kohn J. Evaluation of a series of tyrosine-derived polycarbonates as degradable biomaterials. J Biomed Mater Res. 1994;**28**(8):919–30.

26. Brocchini S, James K, Tangpasuthadol V, Kohn J. A combinatorial approach for polymer design. J Am Chem Soc. 1997 May 1;**119**(19):4553–4.

27. Anderson DG, Peng W, Akinc A *et al*. A polymer library approach to suicide gene therapy for cancer. Proc Natl Acad Sci USA. 2004 Nov 9;**101**(45):16028–33.

28. Anderson D, Tweedie C, Hossain N *et al*. A combinatorial library of photocrosslinkable and degradable materials. Adv Mater. 2006;**18**(19):2614–8.

29. Thaburet JF, Mizomoto H, Bradley M. High-throughput evaluation of the wettability of polymer libraries. Macromol Rapid Commun. 2004;**25**(1):366–70.

30. Meredith JC, Sormana JL, Keselowsky BG, Garcia AJ, Tona A, Karim A *et al*. Combinatorial characterization of cell interactions with polymer surfaces. J Biomed Mater Res. 2003;**66**A(3):483–90.

31. Simon J, Stephens JS, Dorsey SM, Becker ML. Fabrication of combinatorial polymer scaffold libraries. Rev Sci Instrum. 2007 Jul;**78**(7):072207.

32. Liu E, Treiser MD, Patel H *et al*. High-content profiling of cell responsiveness to graded substrates based on combinatorially variant polymers. Combinat Chem High-throughput Screening. 2009;**12**(7):646–55.

33. Yang Y, Bolikal D, Becker ML, Kohn J, Zeiger DN, Simon CG. Combinatorial polymer scaffold libraries for screening cell–biomaterial interactions in 3D. Adv Mater. 2008;**20**(11):2037–43.

34. Bhat RR, Chaney BN, Rowley J, Liebmann-Vinson A, Genzer J. Tailoring cell adhesion using surface-grafted polymer gradient assemblies. Adv Mater. 2005;**17**(23):2802–7.

35. Mei Y, Elliott JT, Smith JR *et al*. Gradient substrate assembly for quantifying cellular response to biomaterials. J Biomed Mater Res. 2006;**79**A(4):974–88.

36. Ionov L, Stamm M, Diez S. Size sorting of protein assemblies using polymeric gradient surfaces. Nano Lett. 2005;**5**(10):1910–4.

37. Iwasaki Y, Sawada S, Nakabayashi N *et al*. The effect of the chemical structure of the phospholipid polymer on fibronectin adsorption and fibroblast adhesion on the gradient phospholipid surface. Biomaterials. 1999;**20**(22):2185–91.

38. Kataoka K, Harada A, Nagasaki Y. Block copolymer micelles for drug delivery: design, characterization and biological significance. Adv Drug Deliv Rev. 2001 Mar 23;**47**(1):113–31.

39. de Loos M, Feringa BL, van Esch JH. Design and application of self-assembled low molecular weight hydrogels. Eur J Org Chem. 2005;3615–32.

40. Sijbesma RP, Beijer FH, Brunsveld L *et al*. Reversible polymers formed from self-complementary monomers using quadruple hydrogen bonding. Science. 1997;**278**(5343):1601–4.

41. Dankers PYW, Hermans TM, Baughman TW *et al*. Hierarchical formation of supra-molecular transient networks in water: a modular injectable delivery system. Adv Mater. 2012;**24**(20):2703–9.

42. Dankers PYW, Harmsen MC, Brouwer LA, Van Luyn MJA, Meijer EW. A modular and supramolecular approach to bioactive scaffolds for tissue engineering. Nat Mater. 2005;**4**(7):568–74.

43. Kieltyka RE, Bastings MMC, Almen GC van *et al*. Modular synthesis of supramolecular ureidopyrimidinone–peptide conjugates using an oxime ligation strategy. Chem Commun. 2012;**48**(10):1452–4.

44. Dankers PYW, van Leeuwen ENM, van Gemert GML *et al*. Chemical and biological properties of supramolecular polymer systems based on oligocaprolactones. Biomaterials. 2006;**27**(32):5490–501.

45. Ellington AD, Szostak JW. In vitro selection of RNA molecules that bind specific ligands. Nature. 1990;**346**(6287):818–22.

46. Zeltinger J, Schmid E, Brandom D *et al*. Advances in the development of coronary stents. Biomater. Forum. 2004;**26**(1),8–9, as cited in Kohn J. New approaches to biomaterials design. Nat Mater. 2004;3(11):745–7.

47. Lehn J-M. Dynamers: dynamic molecular and supramolecular polymers. Prog Polym Sci. 2005 Aug;**30**(8–9):814–31.

48. Corbett PT, Leclaire J, Vial L, West KR *et al*. Dynamic combinatorial chemistry. Chem Rev. 2006;**106**(9):3652–711.

4 Microfabrication techniques in materiomics

Hemant Unadkat[*], Robert Gauvin[*], Clemens A. van Blitterswijk,
Ali Khademhosseini, Jan de Boer and Roman Truckenmüller
[*]Both authors contributed equally

Scope

This chapter deals with an overview of basic microfabrication techniques. The goal is to explain to the reader how such techniques can be utilized in the field of materiomics. The basic processes used in microfabrication including photolithography, etching, electron beam lithography and micromoulding are explained. Some classic examples of these techniques as applied to materiomics are also shown. Furthermore, possible uses of such techniques, and the development and application of hybrid techniques to be able to answer fundamental questions about biological behaviour of materials, are also suggested.

4.1 Basic principles of microfabrication

4.1.1 Introduction

Techniques used to fabricate structures or devices smaller than 100 μm are commonly referred to as microfabrication techniques. Initially meant for the electronics industry, they have found a wide range of applications in diverse fields such as chemical engineering and the life sciences. Since the early 1990s, the application of microfabrication technologies in the area of chemical and biological analysis has been termed 'micro total analysis systems' (μTAS) (1). Microfabricated devices meant for μTAS initially offered the advantage of sample analysis on the microscale, but over the years, the evolution of these technologies has led to the facilitation of sample preparation, fluid handling, separation systems, cell handling and cell culturing in an integrated manner (1). The application of microtechnologies for the fabrication of devices or systems to study material properties benefits from cost efficiency, high performance, precision-based design flexibility, miniaturization and automated analysis. Miniaturization involves the convergence of multiple disciplines, such as fluid dynamics, material sciences, engineering and the life sciences, that need to join expertise in order to design functional systems. Moreover, these devices can be used to evaluate biological behaviour *in vitro* and can

Materiomics: High-Throughput Screening of Biomaterial Properties, ed. Jan de Boer and Clemens van Blitterswijk. Published by Cambridge University Press. © Cambridge University Press 2013.

help us to test thousands of different biomaterials and surface properties without the complexity related to *in vivo* assays.

4.1.2   Materiomics and µTAS

Such µTAS approaches have been adapted to be applied to multiple disciplines such as pharmacology, genetics and proteomics. In pharmacology, for instance, properties of thousands of different compounds are studied using high-throughput screening (HTS) in order to test novel drug candidates. This approach has led to the discovery of new drugs and formulations to treat a given disease or pathology by making it possible to investigate multiple compounds in parallel. Similarly, HTS of gene activity can now be easily monitored using microarray technologies. The new generation of microarrays and HTS systems also provides exceptional possibilities such as fluid handling in integrated devices, greatly improving the reading capabilities and the quality of results. Likewise, materiomics can be used to perform large-scale study of structure, properties and function of natural and synthetic materials.

The use of µTAS in the evolving field of materiomics also provides the possibility of fabricating miniaturized devices that allow the accommodation of thousands of different materials or test conditions on the same platform. This makes it possible to study the behaviour and characteristics of all these materials and test conditions within one experiment.

4.1.3   Microfabrication techniques and processes

Photolithography

The term photolithography refers to transferring geometric patterns into a photosensitive material via selective exposure to light. Figure 4.1 shows an example of a photolithographic process followed by a subtractive pattern transfer by etching (2, 3).

The first step in photolithography typically involves coating a silicon wafer with a photoresist (Figure 4.1b). Two types or tones of photoresist can be used, positive and negative. Here we describe the process using a negative photoresist and an oxidized silicon wafer as an example (Figure 4.1a). A chromium glass photomask, which can be fabricated by laser direct writing, can be used to selectively expose the photoresist with ultraviolet (UV) light (Figure 4.1c). After exposure, the wafer can be immersed in a developer solution to remove the non-cross-linked photoresist, which leaves a pattern of bare and photoresist-coated areas of silicon oxide on the wafer (Figure 4.1d). The areas exposed to UV light remain coated with photoresist, thereby providing a negative image of the mask. The wafer can subsequently be etched using hydrofluoric acid which will remove the bare oxide regions leading to corresponding cavities (Figure 4.1e). The cross-linked photoresist prevents the etchant from reacting with the oxide layer underneath. After etching, the remaining photoresist can be stripped off with a solution such as the acid piranha (H_2SO_4:H_2O_2) which attacks only the photoresist and not the silicon and its oxide (Figure 4.1f). The resolution of photolithography is limited by the wavelength used

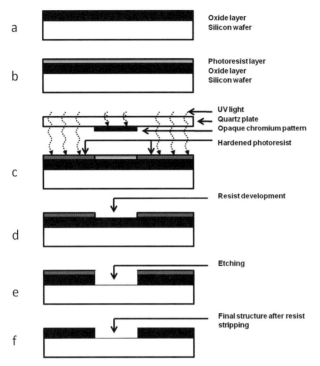

Figure 4.1   Schematic representation of traditional photolithography and subtractive pattern transfer using an oxidized Si wafer.

and can go as low as a few hundred nanometres in contact printing in research and a few tens of nanometres in projection lithography in industry.

Exposure and post-exposure treatment

Microfabrication patterns can be designed using a variety of commercially available software. Different methods are available to fabricate photomasks from the designed patterns, such as the use of laser beam lithography or high-definition printing processes, to selectively expose a photoresist on a substrate of interest.

The resist is applied on the substrate typically by spin-coating. Patterns can be transferred into the resist by shining light through the mask (Figure 4.1c). Usually, different wavelengths of light, e.g. 435 nm (g-line), 405 nm (h-line) and 365 nm (i-line) of a mercury arc lamp are used for exposure of photoresist. The exposure of photoresist to UV light either increases or decreases the solubility of the resist in an appropriate developer depending on whether a positive or negative resist is used. Therefore, in the case of a positive resist, the exposed areas will dissolve during resist development, and vice versa in the case of a negative resist. The profile of the developed photoresist layer depends amongst others on the resist tone, exposure dose, developer strength and development time: a desired side wall profile can be obtained by modifying these parameters. Prior to resist development, post-exposure treatment is applied, typically post-exposure bake, but also flood exposure, a treatment with reactive gas or vacuum treatment.

Development, descumming and post-baking

The process of developing the resist involves its selective dissolving (Figure 4.1d). Typically, negative resists are developed using organic solvents and positive resists using aqueous alkaline solutions such as tetramethyl ammonium hydroxide. Occasionally, resist residues remain entrapped in the pattern after development. A process called descumming, involving mild oxygen plasma, is used to get rid of these residues. In this process, energetic oxygen ions bombard and react with the residual resist and burn it. Then a post-baking step is performed which removes residual solvents and promotes adhesion of the resist film to the substrate beneath. Post-baking, also referred to as hard baking, increases the hardness of the film and its resistance to subsequent etching and deposition steps.

Direct writing techniques

Direct writing techniques are a set of serial techniques. Unlike photolithography, where the whole wafer can be patterned at once, serial writing incrementally patterns multiple small areas of a substrate. Such processes rely on the use of tools such as lasers, electron beams (e-beam), focused ion beams or atomic force microscopes (AFM). These techniques are highly accurate in terms of feature dimensions, and resolutions down to around 10 nm can be achieved.

Direct writing techniques are relatively slow and expensive, and their use is limited to applications such as mask fabrication and patterning of small areas. In a common usage scenario, these techniques can be efficiently used to write areas of a few hundred square micrometres. The resolution of direct writing is limited amongst others by the spot and step size of the corresponding technique and the substrate properties.

Electron beam, focused ion beam and laser-based techniques

These techniques all rely on the use of light or particle beams for which the wavelengths are extremely small, allowing the fabrication of correspondingly small structures in the nanometre range. With the advent of systems including multiple e-beams, this direct writing procedure is becoming increasingly faster. Along with the use of multiple beams, the areas being written on can be stitched together using corresponding software and appropriate stages. The stitching method allows areas in the range of a few square centimetres to be written.

The use of a particle beam to pattern photoresist-coated wafers in selective areas is similar to the previously discussed photolithography technique. Thus, it requires all the subsequent steps used in photolithography including photoresist development, pattern transfer to the substrate and resist stripping. Even though there have been numerous advances with respect to the quality of the beam being used, the corresponding techniques are hampered by limitations due to the atomic or molecular structure of the substrate.

In a second scenario, structures can be directly written into the substrate by means of a focused ion beam. It has been recently shown that resolutions as low as 6 nm can be achieved using helium ions (4).

Figure 4.2   AFM scan of a 25 nm tall rendition (left) of the Matterhorn, a 14 692 foot tall alpine mountain (right) (photographer: Marcel Wiesweg; source: Wikimedia), at a scale of 1-to-5 billion. The model was sculpted using AFM-based direct writing.

Techniques based on atomic force microscopy

AFM, a technique conventionally used in surface and materials characterization, is increasingly used for nanoscale fabrication. For direct-writing techniques, the tip of an AFM can be used like a needle to displace or remove undesired material, thereby forming point- or line-shaped craters. The tip can also be used in order to deposit material just like a conventional ink pen. The latter method is termed dip pen nanolithography (DPN). The former can be used on relatively soft substrates like polymers for creating localized topographic patterns. AFM-based direct patterning techniques are generally considered slow and time-consuming. However, researchers from IBM have recently demonstrated that using a heated cantilever technique improves the throughput of such a technique (5). They demonstrated that it is possible to reproduce a 25 nm high 3D replica of the Matterhorn (Figure 4.2) from a glassy molecular resist in less than three minutes.

Creating nanostructures using DPN is a single-step process which does not require the use of resists. Using a conventional AFM, it is possible to achieve ultra-high-resolution features – as small as 15 nm line widths and approximately 5 nm spatial resolutions. For instance, using DPN, molecules of alkane thiols or proteins which bind strongly to gold can be selectively deposited in a defined format on gold substrates.

Box 4.1 Classic experiment

Using electron beam lithography (EBL), Dalby *et al.* (46) fabricated topographic patterns on a silicon wafer in the form of arrays of 120 nm diameter pits of 100 nm depth and 300 nm pitch in hexagonal and square arrangements. The arrays were also fabricated with near square order of dots, but random displacements of ±20 nm and ±50 nm were introduced maintaining an average 300 nm pitch. Finally, totally random arrangements were fabricated. The silicon wafer was subsequently used to prepare a nickel shim using electroplating. Using the nickel shim as mould, patterns were transferred to poly(methyl methacrylate) (PMMA) blocks using hot embossing.

Human mesenchymal stem cells (hMSC) were cultured on both patterned and non-patterned PMMA substrates, and osteopontin and osteocalcin expression was quantified using immunofluorescence staining (Figure 4.3). It was revealed that cells cultured on the disordered square array with dots displaced randomly by up to 50 nm on both axes (which they call the DSQ-50 surface) resulted in higher expression of osteogenic markers compared with non-patterned substrates.

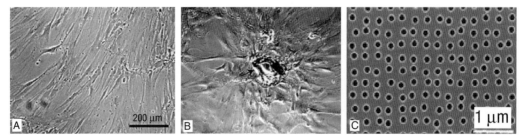

Figure 4.3   Phase-contrast/bright-field images showing (A) fibroblastic morphology of MSCs on the planar control after 28 days in contrast to (A) mature bone nodules containing mineral on displaced square 50 arrangement (±50 nm from true centre) surface (DSQ-50). (C) SEM image of 120 nm pits fabricated by EBL in DSQ-50 arrangement which led to bone nodule formation.

Polymer micromoulding techniques

Many microfabrication techniques use a mould or a master to replicate microstructures into mouldable polymeric materials. Thermoplastic polymers, which are a group of materials that can be reshaped when heated around or above the softening temperature of the material, are frequently used for such processes. A molten thermoplastic polymer, a polymer solution or a thermally or photo-curable polymer resin can be pressed, casted or drawn by capillary action into the mould space and subsequently solidified in it. Sheets from thermoplastic polymers or corresponding layers on wafers can also be embossed or imprinted by heating and thus softening the polymer and then pressing the mould into it. Micromoulds can be engineered using, for example, traditional photolithographic processes. Depending upon the dimensions to be achieved, the moulds can also be fabricated using direct writing processes. Moulding processes are fairly straightforward, fast and inexpensive, and can be easily up-scaled for bulk manufacturing.

Soft lithography

Soft lithography is a term suggested by Whitesides and co-workers and is an umbrella term for micro- or nanopatterning processes which use stamps, moulds or masks made from a soft, elastomeric material, most commonly polydimethylsiloxane (PDMS) (6). Soft lithography has lately gained popularity because of the simplicity of methods such as replica moulding (REM), microtransfer moulding (μTM), micromoulding in capillaries (MIMIC) or solvent assisted micromoulding (SAMIM).

The fabrication of moulds in soft lithography includes pouring the PDMS pre-polymer onto a patterned silicon master and curing it at an elevated temperature to replicate the desired features. PDMS casting can also be used to fabricate multilayered integrated microfluidic arrays with channels for fluid transport applications such as perfusion with cell culture media. The PDMS layers can be bonded to each other or to glass after activation in oxygen plasma.

PDMS moulds can be used for patterning hydrogels and biomaterials such as agarose, gelatin and polyethylene glycol (7). Hydrogel-based well arrays can subsequently be used for high-throughput screening of extracellular matrices (ECM) and combinatorial chemistry arrays. Different combinations of ECM components can be spotted into the wells of the hydrogel arrays using an automated spotter. Subsequently, cells can be cultured on these substrates and the interactions with the ECM molecules can be analysed using high-throughput imaging tools (8).

Hot embossing

Embossing uses substrates in the form of sheets of thermoplastic materials or substrates coated with corresponding layers which are patterned using a mould by applying heat and pressure. Thermoplastics routinely used for hot embossing include PMMA, poly(lactic acid) (PLA), polycarbonate (PC), cyclic olefin copolymer (COC), polystyrene (PS), polyvinyl chloride (PVC) and polyethylene terephthalate glycol (PETG).

Upon fabrication of the mould, the selected thermoplastic substrate is placed into a heated press. Here, the substrate is typically sandwiched between the mould and a counter plate. After heating, the sandwich is compressed to emboss the mould into the thermoplastic sheet or layer. One mould can be used for batch fabrication of multiple devices. Some versions of embossing presses are accompanied with a vacuum system to eliminate air bubbles being entrapped between substrate and stamp. The replication fidelity of embossing is mainly limited by the mould fabrication process used.

Microthermoforming

Thermoforming refers to the shaping of a heated semi-finished product in the form of a thermoplastic polymer film (or plate) by three-dimensional stretching it typically into or over a mould. In thermoforming, the film is clamped at its edges or around the mould to be used. The stretching results in thinning of the product compared to its initial thickness.

In micro-pressure forming (Figure 4.4), a thin thermoplastic film is inserted into a microthermoforming tool. The three-part tool consists of a plate-shaped micromould with mould cavities, a counter plate with openings for evacuation and gas pressurization, and an axial seal ring in-between. The tool is mounted into a heated press and closed to such an extent that vacuum sealing of the volume enclosed by the two tool plates is achieved. Then the entire tool is evacuated (Figure 4.4A), completely closed in order to clamp the film and heated to the softening temperature of the polymer. The film is moulded into the evacuated microcavities by using compressed nitrogen (Figure 4.4B).

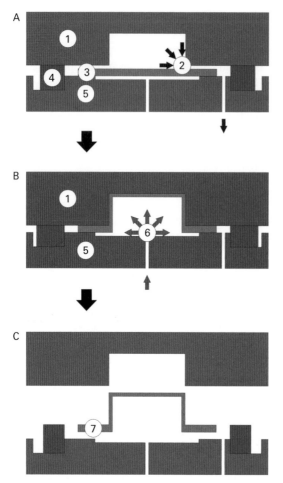

Figure 4.4   Micro-pressure forming with the following process steps: insertion of a thermoplastic film, (A) evacuation of the tool, heating up the tool, (B) forming of the film by compressed nitrogen, cooling down the tool, releasing the nitrogen pressure, and (C) demoulding and unloading the film microstructure (1: mould, 2: vacuum, 3: thermoplastic film, 4: seal, 5: counter plate, 6: compressed nitrogen, 7: thermoformed film microstructure.

The tool is then cooled down, the nitrogen pressure released, the tool opened, and the thermoformed film microstructure demoulded (Figure 4.4C) and unloaded.

Micro injection moulding

Unlike hot embossing and microthermoforming, injection moulding does not use a softened or molten polymer sheet. The process involves molten polymer granules and is usually employed for the fabrication of 3D parts or for industrial-scale production of devices. In an injection moulding process, a sealable mould cavity is fabricated. The mould is equipped with a nozzle for injection of the molten polymer into the void space. Upon filling of the mould, the assembly is cooled down which results in the solidification of the polymer melt. The assembly is subsequently opened and the fabricated parts are ejected.

4.2   The application of microfabrication in materiomics

The microfabrication techniques so far explained can be used to fabricate devices for high-throughput materiomics studies. In addition, efforts have recently been dedicated towards the development of metamaterials using microfabrication techniques. These classes of materials are artificially engineered to display properties which may not be inherent in nature. For instance, metamaterials has provided us with materials having negative refractive index, a property which is not found in natural materials (9). Similarly, materials can be engineered with desirable properties for biomedical applications, such as anti-fouling. Such classes of materials gain their property by virtue of their structure rather than composition. For example, techniques such as AFM-based direct writing can be used for fabrication of a wide variety of 3D metamaterials. Since microscale technologies are currently being investigated as potential tools for addressing biomedical challenges, the engineering of structures and devices such as microsystems issued from materiomics studies will help control and monitor biological phenomena occurring at the microscale.

Box 4.2 Classic experiment

Recently, we have used hot embossing to fabricate a surface-topographic library for analysing the effect of surface topographies on cell behaviour. The device, termed a TopoChip (Figure 4.5), contains 2178 different, randomly generated surface topographies (11). The topographies were designed using a mathematical algorithm. A chromium mask was fabricated and used for patterning resist on a silicon wafer by conventional photolithography, followed by deep reactive ion etching of the topographies into the wafer. Using the silicon master, hot embossing was performed on poly(D,L-lactic acid) sheets in a nanoimprint lithography machine.

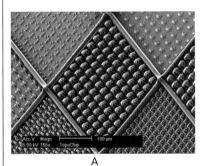
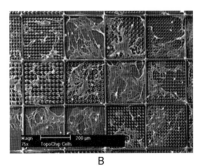

A   B

Figure 4.5 A, SEM of a section of the TopoChip. B, SEM of human mesenchymal stem cells exhibiting diverse morphologies when cultured on the TopoChip.

Proliferation and differentiation potential of surface topographies was evaluated by culturing hMSCs. Topographies with mitogenic potential and with osteogenic differentiation potential were identified by high-content imaging and extensive data mining.

4.2.1   Microfabrication as tool in high-throughput screening of material libraries

Microtechnologies can be used in combination with biocompatible and biodegradable materials to generate material libraries with defined cell scale features to study cell-material interactions in a high-throughput fashion (10, 47–50). For example, the 2×2 cm^2 TopoChip has allowed us to demonstrate that surface topography can influence the proliferation and differentiation of human mesenchymal stromal cells (hMSCs), using a library of randomly designed surface profiles (11). Moreover, using microfabrication and microfluidics strategies, it is also possible to create arrays of multiple materials, proteins, chemicals or stiffness gradients with high resolution and spatial control (12). Therefore, materiomics applications can be used to design and explore new generations of biomaterials with tailored properties. These should allow the development of new technologies such as the engineering of devices and functional biological units that will be investigated as potential solutions to clinical applications (13).

Materiomics and combinatorial approaches have greatly improved the rational design and the development of new classes of biomaterials. At the fundamental level, quantitative analysis of large sample sizes of micro-engineered materials has helped to identify the most efficient designs for defined purposes (14, 15). These platforms have also provided high-throughput assays allowing the measurement of material properties such as wettability (indicating hydrophobicity or hydrophilicity of a sample, measured by contact angle), surface topography, surface chemistry and substrate stiffness (10, 16). This methodology has helped to highlight the influence of different ECM molecules, binding receptors such as integrins, and biomaterials geometry on cell behaviour, spreading and differentiation (17). A micropatterning technique has also been used to control cell shape and spreading with single-cell precision, for thousands of cells at a time. This technique led to the identification of cell shape as a key regulator in mechano-osteoblastic differentiation (18), and therefore contributed to understanding the structure–function relationship between material properties and biological performance (12). Miniaturization has also proved to be efficient in the development of biologically relevant high-throughput assays, which resulted in multiple ground-breaking studies on fundamental cell behaviour that would not previously have been possible (19, 20).

Microfabricated microarrays greatly facilitate the rapid synthesis of libraries, and high-throughput analysis enables the assessment of multiple conditions, providing a general framework for the combinatorial development of synthetic substrates for biomedical applications. Results obtained from these assays can then be used as a starting point to enhance the design of biomaterials for tissue engineering and regenerative medicine.

This approach, which radically differs from previous strategies that relied on the development of, for example, a single polymer on which multiple experiments were conducted, represents a much more efficient way to develop and tailor the properties of new materials. In a recent study, a high-throughput analysis was performed using a library of 50 000 compounds to find the best substrate in order to promote self-renewal of mouse embryonic stem cells, greatly reducing the amount of time and effort required to obtain the optimal result (21).

4.2.2   Microfabrication techniques for scaffold fabrication

Photolithography and soft lithography have also considerably improved the control over the microarchitecture of scaffolding materials (22). These fabrication techniques have proved to be successful in controlling mechanical and chemical properties, interconnectivity, geometry and isotropy of biodegradable polymers with micrometre-scale resolution (22). Microfabricated substrates were also used to deliver drugs in a precisely controlled fashion by engineering the porosity and the crosslinking density of the carrier used for drug transport (23). Microfabrication techniques have been applied for the preparation of multiple biodegradable polymer scaffolds, e.g. poly(L-lactic acid) (PLLA), poly(lactic-co-glycolic acid) (PLGA) and poly(glycerol sebacate) (PGS), and were used to produce hydrogel-based scaffolds with tunable transport properties, resulting in improved outcomes (24). These techniques were also used to engineer surface topographies for guiding cell adhesion, orientation and migration (25, 26). These approaches are contributing greatly to efforts in the field of tissue engineering to reproduce cell–cell and cell–ECM interactions in engineered tissues. They can also be used to control the spatial distribution of molecules and cells, and to create physiologically relevant gradients in biomaterials (27). These physical, chemical and biological cues can later on be used to control cell adhesion, migration and proliferation and improve biological function of these systems.

Bottom-up approaches using organ printing or projection stereolithography are also part of the effort to generate metamaterials that could lead to new ways of engineering living tissues or incorporating vascular structures into 3D constructs (28, 29). Furthermore, innovative approaches combining microfabrication of building blocks and the bottom-up mesoscale assembly of these components to scale up 3D structures can be used in synergy to produce functional 3D structures *in vitro* (30–33). Thus, these technologies can help to provide significant insight into cell behaviour *in vitro* and can result into great technological advances *in vivo*.

4.3   Future perspectives

Application of materiomics and microfabrication to biotechnology has aided the rapid expansion of various research fields such as cellular and molecular assays, diagnostic devices, drug discovery and chemical and biological detection (6, 34–38). New methods are moving towards robotic nanotechnology to deliver nanolitre volumes of many different molecules or materials. As the size of devices decreases, their surface-to-volume ratio increases. For this reason, the surface properties become very important in determining the performance of the assay. Therefore, it is necessary to engineer surface properties with molecular-level precision and to build platforms facilitating the reading of the assays. The combination of microfluidics with photopolymerization chemistry has recently resulted into hydrogels containing gradients of cross-linking and signalling or adhesive molecules across the material, thus resulting in the regulation of cell behaviour such as migration, adhesion and differentiation within the gel (39–41). These technologies could help to closely control the restoration of tissue morphology and function since they can be used to control the area, shape and locations of the substrate on which cells attach. They also have tremendous potential in overcoming key challenges such as engineering a microvasculature, as well as introducing complexity in engineered tissues (42). Based on materiomics and microscale technologies, a large set of tools are available to investigate cell–cell and cell–microenvironment interactions using high-throughput technologies and to test many environmental factors simultaneously.

The application of combinatorial approaches could also lead to new ways to engineer biomimetic 3D constructs (43–45). For example, one could use materiomics studies to precisely engineer the properties of building blocks to build 3D functional organs using a mesoscale assembly approach. These building blocks could comprise precise physical, biological or chemical cues or gradients that would promote tissue regeneration.

Future challenges confronting tissue engineers will include the design of novel 3D matrices that can precisely control the cellular microenvironment. These materials will have to incorporate specific ligands, tailored mechanical cues and controlled release of growth factors. The development of these 'smart' matrices means giving the cells the appropriate signals in order to induce adhesion, migration, proliferation or differentiation, depending on whether the tissue is in a repair, regeneration or remodelling phase. Microfabrication and materiomics are already powerful tools for basic discoveries, but there is still a need for the implementation of currently available systems and for the combination of this expertise with other disciplines to produce therapeutic outcomes for clinical applications and improve the efficiency of diagnostic devices.

4.4   Snapshot summary

- The physical structure of biomaterial interface is as important as the chemical interactions. Physical properties such as surface area, wettability and topography have lately been found to influence the outcome of surgical procedures.
- The physical properties such as the topography of a biomaterial surface can be changed by various methods such as sand blasting, salt leaching and so on. However, the surface modifications resulting from these methods are not well controlled.
- Microfabrication provides numerous methods to change the surface characteristics of biomaterials in a reliable, reproducible and convenient manner down to nanometre precision.

- The use of microfabrication techniques in materiomics can be divided into two classes. On one hand they can be used for miniaturizing assays of materials, i.e. development of devices for high-throughput screening of biomaterials, and on the other hand they can be used for fabrication of metamaterials.
- Conventional microfabrication techniques such as photolithography and etching have been explained in sufficient detail.
- Advanced techniques such as electron beam lithography, dip pen lithography and direct AFM-based writing have also been explained and examples given.
- In addition, replica moulding techniques such as micro injection micromoulding, soft lithography, nanoimprint lithography and microthermoforming have been explained.
- The adaptability of these techniques is already leading to high-throughput libraries of biomaterials.

Further reading

Gobaa S, Hoehnel S, Roccio M *et al*. Artificial niche microarrays for probing single stem cell fate in high-throughput. Nat Meth. 2011;**8**(11):949–55.

Unadkat HV, Hulsman M, Cornelissen K *et al*. An algorithm-based topographical biomaterials library to instruct cell fate. Proc Natl Acad Sci USA. 2011;**108**(40):16565–70.

Anderson DG, Levenberg S, Langer R. Nanolitre-scale synthesis of arrayed biomaterials and application to human embryonic stem cells. Nat Biotechnol. 2004;**22**(7):863–6.

McBeath R, Pirone DM, Nelson CM, Bhadriraju K, Chen CS. Cell shape, cytoskeletal tension, and RhoA regulate stem cell lineage commitment. Dev Cell. 2004;**6**(4):483–95.

Gauvin R, Chen Y-C, Lee JW *et al*. Microfabrication of complex porous tissue engineering scaffolds using 3D projection stereolithography. Biomaterials. 2012;**33**(15):3824–34.

Dalby MJ, Gadegaard N, Tare R *et al*. The control of human mesenchymal cell differentiation using nanoscale symmetry and disorder. Nat Mater. 2007;**6**(12):997–1003.

References

1. Andersson H, Berg A. *Lab-On-Chips for Cellomics: Micro and Nanotechnologies for Life Science*: Kluwer Academic; 2004.
2. Gad-el-Hak M. *MEMS: Introduction and Fundamentals*: CRC/Taylor & Francis; 2006.
3. Madou MJ. *Fundamentals of Microfabrication : The Science of Miniaturization* 2nd edn: CRC; 2002.
4. Maas D, van Veldhoven E, Chen P *et al*. Nanofabrication with a helium ion microscope. Metrol Inspect Proc Control Microlithog. 2010;**XXIV**:7638.
5. Pires D, Hedrick JL, De Silva A *et al*. Nanoscale three-dimensional patterning of molecular resists by scanning probes. Science. 2010;**328**(5979):732–5.
6. Whitesides GM, Ostuni E, Takayama S, Jiang X, Ingber DE. Soft lithography in biology and biochemistry. Annu Rev Biomed Eng. 2001;**3**:335–73.
7. Kobel S, Limacher M, Gobaa S, Laroche T, Lutolf MP. Micropatterning of hydrogels by soft embossing. Langmuir. 2009;**25**(15):8774–9.

8. Gobaa S, Hoehnel S, Roccio M, Negro A, Kobel S, Lutolf MP. Artificial niche microarrays for probing single stem cell fate in high-throughput. Nat Meth. 2011;**8**(11):949–55.

9. Padilla WJ, Basov DN, Smith DR. Negative refractive index metamaterials. Mater Today. 2006; **9**(7–8):28–35.

10. Underhill GH, Bhatia SN. High-throughput analysis of signals regulating stem cell fate and function. Curr Opin Chem Biol. 2007;**11**(4):357–66.

11. Unadkat HV, Hulsman M, Cornelissen K *et al.* An algorithm-based topographical biomaterials library to instruct cell fate. Proc Natl Acad Sci. 2011;**108**(40):16565–70.

12. Davies MC, Alexander MR, Hook AL *et al.* High-throughput surface characterization: A review of a new tool for screening prospective biomedical material arrays. J Drug Target. **18**(10):741–51.

13. Mei Y, Saha K, Bogatyrev SR *et al.* Combinatorial development of biomaterials for clonal growth of human pluripotent stem cells. Nat Mater. 2010;**9**(9):768–78.

14. Anderson DG, Levenberg S, Langer R. Nanolitre-scale synthesis of arrayed biomaterials and application to human embryonic stem cells. Nat Biotechnol. 2004;**22**(7):863–6.

15. Sundberg SA. High-throughput and ultra-high-throughput screening: solution- and cell-based approaches. Curr Opin Biotechnol. 2000;**11**(1):47–53.

16. Mei Y, Hollister-Lock J, Bogatyrev SR *et al.* A high-throughput micro-array system of polymer surfaces for the manipulation of primary pancreatic islet cells. Biomaterials. **31**(34):8989–95.

17. Keselowsky BG, García AJ. Quantitative methods for analysis of integrin binding and focal adhesion formation on biomaterial surfaces. Biomaterials. 2005;**26**(4):413–8.

18. McBeath R, Pirone DM, Nelson CM, Bhadriraju K, Chen CS. Cell shape, cytoskeletal tension, and RhoA regulate stem cell lineage commitment. Dev Cell. 2004;**6**(4):483–95. Epub 2004/04/08.

19. Flaim CJ, Chien S, Bhatia SN. An extracellular matrix microarray for probing cellular differentiation. Nat Methods. 2005;**2**(2):119–25.

20. Soen Y, Mori A, Palmer TD, Brown PO. Exploring the regulation of human neural precursor cell differentiation using arrays of signaling microenvironments. Mol Syst Biol. 2006;**2**:37.

21. Chen S, Do JT, Zhang Q *et al.* Self-renewal of embryonic stem cells by a small molecule. Proc Natl Acad Sci USA. 2006;**103**(46):17266–71.

22. Khademhosseini A, Langer R, Borenstein J, Vacanti JP. Microscale technologies for tissue engineering and biology. Proc Natl Acad Sci USA. 2006;**103**(8):2480–7.

23. Richards Grayson AC, Choi IS *et al.* Multi-pulse drug delivery from a resorbable polymeric microchip device. Nat Mater. 2003;**2**(11):767–72.

24. Khademhosseini A, Langer R. Microengineered hydrogels for tissue engineering. Biomaterials. 2007;**28**(34):5087–92.

25. Guillemette MD, Cui B, Roy E *et al.* Surface topography induces 3D self-orientation of cells and extracellular matrix resulting in improved tissue function. Integr Biol (Camb). 2009;**1**(2):196–204.

26. Alaerts JA, De Cupere VM, Moser S, Van den Bosh de Aguilar P, Rouxhet PG. Surface characterization of poly(methyl methacrylate) microgrooved for contact guidance of mammalian cells. Biomaterials. 2001;**22**(12):1635–42.

27. Hancock MJ, He J, Mano JF, Khademhosseini A. Surface-tension-driven gradient generation in a fluid stripe for bench-top and microwell applications. Small. 2011;**7**(7):892–901. Epub 2011/03/05.

28. Gauvin R, Chen Y-C, Lee JW *et al*. Microfabrication of complex porous tissue engineering scaffolds using 3D projection stereolithography. Biomaterials. 2012;**33**(15):3824–34.

29. Guillemot F, Souquet A, Catros S *et al*. High-throughput laser printing of cells and biomaterials for tissue engineering. Acta Biomater. 2010;**6**(7):2494–500.

30. Bowden N, Terfort A, Carbeck J, Whitesides GM. Self-assembly of mesoscale objects into ordered two-dimensional arrays. Science. 1997;**276**(5310):233–5.

31. Du Y, Lo E, Ali S, Khademhosseini A. Directed assembly of cell-laden microgels for fabrication of 3D tissue constructs. Proc Natl Acad Sci. 2008;**105**(28):9522–7.

32. Fernandez JG, Khademhosseini A. Micro-masonry: Construction of 3D structures by microscale self-assembly. Adv Mater. 2010;**22**(23):2538–41.

33. Gauvin R, Khademhosseini A. Microscale technologies and modular approaches for tissue engineering: Moving toward the fabrication of complex functional structures. ACS Nano. 2011;**5**(6):4258–64.

34. Foquet M, Korlach J, Zipfel W, Webb WW, Craighead HG. DNA fragment sizing by single molecule detection in submicrometer-sized closed fluidic channels. Analyt Chem. 2002;**74**(6):1415–22.

35. Kennedy GC, Matsuzaki H, Dong S *et al*. Large-scale genotyping of complex DNA. Nat Biotechnol. 2003;**21**(10):1233–7.

36. Lipshutz RJ, Fodor SP, Gingeras TR, Lockhart DJ. High density synthetic oligonucleotide arrays. Nat Genet. 1999;**21**(1 Suppl):20–4.

37. Sia SK, Linder V, Parviz BA, Siegel A, Whitesides GM. An integrated approach to a portable and low-cost immunoassay for resource-poor settings. Angew Chem Int Ed Engl. 2004;**43**(4):498–502.

38. Rossier JS, Girault HH. Enzyme linked immunosorbent assay on a microchip with electrochemical detection. Lab Chip. 2001;**1**(2):153–7.

39. Burdick JA, Khademhosseini A, Langer R. Fabrication of gradient hydrogels using a microfluidics/photopolymerization process. Langmuir. 2004;**20**(13):5153–6.

40. King KR, Wang CC J, Kaazempur-Mofrad MR, Vacanti JP, Borenstein JT. Biodegradable microfluidics. Adv Mater. 2004;**16**(22):2007–12.

41. Bettinger CJ, Weinberg EJ, Kulig KM *et al*. Three-dimensional microfluidic tissue-engineering scaffolds using a flexible biodegradable polymer. Adv Mater. 2006;**18**(2):165–9.

42. Ingber DE, Mow VC, Butler D *et al*. Tissue engineering and developmental biology: going biomimetic. Tissue Eng. 2006;**12**(12):3265–83.

43. Kaihara S, Borenstein J, Koka R *et al*. Silicon micromachining to tissue engineer branched vascular channels for liver fabrication. Tissue Eng. 2000;**6**(2):105–17.

44. Borenstein JT, Terai H, King KR *et al*. Microfabrication technology for vascularized tissue engineering. Biomed Microdevices. 2002;**4**(3):167–75.

45. Huh D, Matthews BD, Mammoto A *et al*. Reconstituting organ-level lung functions on a chip. Science. 2010;**328**(5986):1662–8.

46. Dalby MJ, Gadegaard N, Tare R *et al*. The control of human mesenchymal cell differentiation using nanoscale symmetry and disorder. Nat Mater. 2007;**6**(12):997–1003.

47. Markert LD, Lovmand J, Foss M *et al*. Identification of distinct topographical surface microstructures favoring either undifferentiated expansion or differentiation of murine embryonic stem cells. Stem Cells Dev. 2009;**18**(9):1331–42.

48. Lovmand J, Justesen J, Foss M *et al*. The use of combinatorial topographical libraries for the screening of enhanced osteogenic expression and mineralization. Biomaterials. 2009;**30**(11):2015–22.

49. Papenburg BJ, Vogelaar L, Bolhuis-Versteeg LA *et al.* One-step fabrication of porous micropatterned scaffolds to control cell behavior. Biomaterials. 2007;**28**(11):1998–2009. Epub 2007 Jan 18.
50. Truckenmüller R, Giselbrecht S, Escalante-Marun M *et al.* Fabrication of cell container arrays with overlaid surface topographies. Biomed Microdevices. 2012;**14**(1):95–107.

5 Bioassay development

Hugo Fernandes, Roderick Beijersbergen, Lino Ferreira, Koen Dechering,
Prabhas Moghe and Katharina Maniura-Weber

Scope

Combinatorial chemistry and high-throughput synthesis of novel materials warrant a paradigm shift in current methods to analyse biological responses. This chapter will provide an overview on bioassay development and how novel assays amenable to high-throughput screening platforms can be adapted to more complex systems. Special emphasis will be devoted to the development of assays that can be used in platforms that closely mimic the *in vivo* complexity of tissues and organs. In that respect, assays that can cope with co-culture systems as well as 3D environments will be discussed. Moreover, modifications or development of new assays and techniques will be described as well as their respective advantages and disadvantages.

5.1 Basic principles of assay development

The ability to measure the speed of light changed the field of physics and the world. Chemical reactions led to the Big Bang and the creation of the Universe, but the ability to measure and control those reactions changed the face of the Earth. We can surely say that the need to see more, and in more detail, led to the development of technologies that made that possible and ultimately contributed to the advance of science and society.

In the eighteenth century, Antoni van Leeuwenhoek was the first to see organisms and cells under a microscope that he built. This observation paved the way to the high-resolution microscopes we use today which allow us to see strands of DNA and other nanoscopic components in cells such as single molecules. Despite the rapid development of van Leeuwenhoek's microscope, measuring techniques lagged for many years until their implementation in today's laboratories.

Monitoring cells and cellular activity *in vitro* and *in vivo* allows us to identify cellular processes characteristic of normal and/or abnormal development such as those occurring in cancer. Numerous cell and non-cell based assays have been developed during the past decades, driving our understanding of biological processes. For example, the development of monoclonal antibodies gave us the possibility to identify specific proteins in complex tissues (1, 2). The identification of cellular components involved in intracellular

Materiomics: High-Throughput Screening of Biomaterial Properties, ed. Jan de Boer and Clemens van Blitterswijk. Published by Cambridge University Press. © Cambridge University Press 2013.

signalling allowed scientists to devise assays to measure the different inhibitors/ enhancers in a signalling cascade, and to design assays to measure their activity as well as approaches to modify such pathways.

Many studies have shown us that the first and most important components in developing an assay are a thorough understanding and characterization of what to measure. We must then design a testing system that allows us to control one or more parameters at will. What can be measured is conditioned by a plethora of factors, some of which are the equipment, budget and staff available, time and feasibility. Often the decision will be made between a biochemical and/or a cell-based assay. The former can be seen as a quick and useful way to analyse a specific target using a reductionist approach, whereas the latter aims to mimic the complexity of the cellular environment or even *in vivo* systems. Some examples of biochemical assays are enzymatic activity of alkaline phosphatase (a protein known to be involved in bone formation), or fibroblast growth factor receptor (FGFR) kinase activity (a receptor involved in bone development) (6, 7). These assays are usually specific for the given target regardless of the complexity of the overall biochemical process, e.g. regulation of alkaline phosphatase or receptor activation. Cell-based assays such as reporter-based assays take into account the regulation of a given signalling pathway by other cellular players, and their readout is the combination of cellular reactions leading to activation of a certain gene, for example. Finally, readouts based on cellular proliferation and survival capture the complex interplay of many cellular processes, often complicating the interpretation of such screens. Most of the assays mentioned above were developed and implemented simultaneously in academic laboratories as well as in the pharmaceutical industry. However, the capacity of pharmaceutical companies to design and synthesize thousands of different drugs for a given target drove the development of new technologies to analyse their role in biological systems. Automation and miniaturization were the solutions devised by the pharmaceutical industry to tackle this problem, which led to the use of the term high-throughput screening (HTS) (8).

The development of HTS assays was mostly done using tissue culture plates, at first using the (now laboratory standard) 96-well plates but very quickly moving to 384 and even 1536 wells per plate. Every step towards higher numbers warrants several optimization steps. For example, the number of cells per assay, maintenance of cells during the total period of the assay, plate effects such as evaporation on the sides and adaptation of fluidic systems able to cope with the ever-smaller volumes needed are just some of the aspects to consider when downscaling the assays. Pushed by pharmaceutical companies, HTS assays became more user-friendly, opening the possibility of establishing such assays in academic laboratories. A few years ago, observing and measuring a few cells posed a great challenge. Today, HTS allows us to analyse thousands of individual cells per assay at a subcellular level in a very short time span. For example, Figure 5.1 shows an example of assays that allows the detection of drugs mimicking hypoxia.

In the past, most laboratories would screen a few compounds but today, adaptation of this assay allows us to screen thousands of compounds in the same period of time (9). However, in contrast to the pharmaceutical industry in which biochemical and cell-based assays were the standard and therefore performed in tissue culture plates, academia had different needs. One particular field that has seen rapid development since the 1990s is

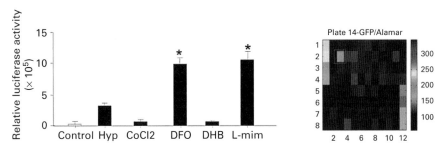

Figure 5.1   Reporter assay for compounds that mimic hypoxia. On the left graph five compounds (black bars) were tested and compared with a control situation (white bar), whereas on the right, using a similar assay optimized for a 96-well plate, 80 compounds were tested in one single assay. Adapted from (9). Copyright (2008) National Academy of Sciences USA.

the field of tissue engineering. Here, cells and growth factors are combined with a carrier material that initially served solely as a delivery system but very quickly evolved to become the focus of attention as an instructive entity capable of controlling cell fate (10, 11). This new discipline created a new need: assays where cells had to be cultured on different platforms/substrates instead of the classic tissue culture plates, therefore posing new challenges for assay development. The new platforms brought with them several issues: chemically variable substrates that can influence how cells adhere and grow; topographical cues that can control cell fate but pose a challenge for imaging-based assays due to differences in focal plane; more time-demanding assays given that the effect of those platforms on cell fate is likely to occur over a long period of time instead of minutes as in some biochemical assays (12–17).

In addition to the complexity of biological systems, then, we are now faced with an extra level of complexity posed by the platform itself. Nevertheless, there are some important guidelines that need to be considered regardless of the platform used.

1. Which biological process do you want to analyse?
2. What type of assay do you need or can be implemented?
3. How robust is the developed assay?
4. How can we deselect false positives that are inevitably part of any large-scale screening dataset?
5. How much intra- and inter-assay variation is expected?
6. Which follow-up assays can be performed?
7. What are the costs?

The choice of the *biological process* to be questioned is the most critical aspect of any assay. What you want to measure is conditioned by your research question. Therefore, you need to define the target clearly prior to development of an assay. Once the target is defined one can move to the *type of assay* and start the optimization process. A few considerations to account for at this stage are the technologies available as well as the biological material needed in order to perform the assay. For example, if analysing enzyme–substrate interactions we need to ensure we can isolate both in sufficient quantities. When the choice is a cell-based assay we need to consider whether primary cells or cell lines should be used, and consider their pros and contras such as heterogeneity, genomic stability, cell survival and proliferation, donor variability, access to the cell source and even ethical issues (18, 19). The *robustness* of the assay can be analysed using statistical methods such as the signal-to-noise (S–N) and the signal-to-background (S–B) ratios and the Z' factor. These metrics will allow the user to discriminate a true signal from a false positive or background noise due to stochastic events. The S–N, S–B ratio and the Z' factor can be calculated as follows:

$$S{-}N = \frac{mean\ signal}{mean\ background}$$

$$S{-}B = \frac{(mean\ signal - mean\ background)}{standard\ deviation\ background}$$

$$Z'\ factor = 1 - \frac{3(standard\ deviation\ signal + standard\ background)}{|mean\ signal - mean\ background|}$$

An ideal assay will have a Z' factor $= 1$ whereas a Z' factor < 0 will not separate the signal from the background and therefore should not be accepted for HTS assays (usually a Z' factor >0.5 is considered amenable to HTS) (20–22).

The *intra- and inter-assay variation* depends on several parameters such as edge effects due to faster evaporation in the edge of the plates, variation in cell culture conditions (for example serum batch used) or in the reagents used for the isolation of your target protein, cell handling during the assay, cell state (in which stage of the cell cycle are the cells when performing the assay), genomic stability during the course of the experiment (do the test compounds lead to mutations and changes in chromosome number?), the passage number of the cells used (it is known that certain phenotypes are lost during extensive culture) and, last but not least, the donor variation issue (what makes each of us unique will also make every assay unique). It is important to distinguish between technical and biological replicates when performing cell-based assays (23). In general, the largest variation is observed with independent biological replicates. The requirement of multiple replicates also necessitates the use of controls for normalization and analysis. The randomization of the controls throughout the plate can ameliorate some, but not all, of the variation and a per-plate control is an absolute requirement for cell-based assays (20). Upon optimization of the mentioned parameters and after the assay we will end up with hits that need further confirmation and validation. Therefore, it is important that *follow-up assays* (secondary assays) are in place to validate the hits. Dose–response experiments and complementary assays are usually performed to validate the hits. Last but not least, we are faced with *costs* during the development of HTS. From the equipment necessary to handle such large numbers of samples to the computer power necessary for the analysis, everything needs to be considered.

Assay development is certainly the most time-consuming part of the process but also the one that, if not done properly, will cause major problems at a later stage. The pharmaceutical industry has optimized several assays over the years, but in the new field of materiomics we need to start almost from the beginning in order to adapt those assays to the new platforms. Figure 5.2 shows an overview of the workflow during assay development.

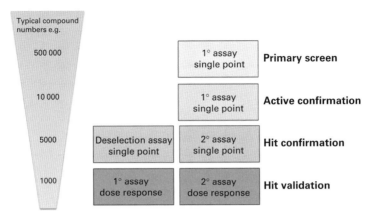

Figure 5.2   The HTS campaign starts with a primary assay in which compounds are tested at a single concentration (typically between 1 and 10 μM) to allow for high throughput. Compounds that are active in this assay are retested in the same assay (in a process called active confirmation) to eliminate false positives that randomly occur in large screening sets. The aim of the next stage (hit confirmation) is to eliminate false positives that show a signal in the primary assay by interfering with the assay technology rather than by modulating the target or signalling pathway of interest. To this end, compounds are tested in a 'deselection' assay that uses the same technology on a non-related target. In addition, compounds are tested in a secondary assay that uses different technology to monitor activity of the same target as assayed in the primary assays. Compounds that pass the deselection and secondary assays are then tested in a range of concentrations to establish the potency of the compound, in a process called hit validation.

Box 5.1 Classic experiment

High-throughput screening of biomaterials and natural extracellular matrix (ECM) components can be used to evaluate the influence of various physico-chemical signals on cellular behaviour. Exploring these combinations enables the screening of cellular responses such as cell viability, proliferation and differentiation. The high-throughput screening of cell–biomaterial interactions can be useful for a number of applications such as drug discovery, chemical toxicity testing and tissue engineering (5). Typically the readout of these cellular assays is based on the measurement of a single end-point per spot.

   Bhatia and collaborators have reported one of the first bioassays for the evaluation of stem cell microenvironment (Figure 5.3) (4). A robotic spotting technology was used to deposit ECM components on a standard microscope slide coated with poly (acrylamide) gel (80 μm thickness). Mixtures of collagen I, collagen III, collagen IV, laminin and fibronectin were used to recreate microenvironments for stem cell differentiation. ECM mixtures (1–2 nL) were deposited per 150 μm diameter spot on the acrylamide gel. The ECM mixtures were retained in the poly(acrylamide) gel by hydrophobic interactions. These spots can accommodate more than 20 cells. The non-fouling properties of the poly(acrylamide) gel prevented cell migration over

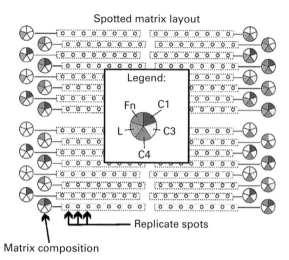

Spotted matrix layout

Legend:

Fn   C1

L   C3

C4

Replicate spots

Matrix composition

Figure 5.3 Schematic representation of the different combinations of extracellular matrix proteins used to create a screening platform capable of controlling cell fate. Adapted with permission from (4).

relatively long periods of time. According to the authors, 20 ECM microarrays could be produced simultaneously in 1 h. The slides maintained spatially confined cellular islands for at least 7 days.

Primary rat hepatocytes were initially tested, chosen because the authors were interested in identifying a hepatic microenvironment. Cell morphology and viability assays were performed to demonstrate cell adhesion on spotted regions and survival. The effects of the ECM composition was evaluated by immunofluorescence for intracellular albumin (a marker of liver specific function) and analysed on days 1 and 7. The results show that the 15 highest albumin signals were associated with underlying matrices containing collagen IV. These studies were extended to mouse embryonic stem cells, using the reporter cell line I114. *In vitro* differentiation of I114 cell line induces reporter (β-galactosidase) expression that coincides with endodermal gene upregulation and colocalization with α-fetoprotein and albumin protein. The cells were cultured in the presence of leukaemia inhibitory factor (1000 units/mL) or retinoic acid (10^{-6} M). The results obtained showed that matrices with collagen I had the highest signal.

5.2   Relation to materiomics

To develop new assays in the field of materiomics we need to take into account the enormous variation in physico-chemical parameters of the platforms/materials used to create them, very similar to ones used in the pharmaceutical industry to generate thousands of different chemical structures (compounds) for a certain target. Ultimately, some of these physico-chemical parameters or combinations of those can control how cells react and hence control their differentiation into different lineages, how proteins bind to different surfaces therefore dictating the integration of the surface with target tissues, how the material will behave on *in vivo* implantation in a 'hostile' environment and so forth. Novel assays to predict the outcome of a certain material or combination of materials and analyse how physico-chemical parameters affect cell fate and tissue formation/remodelling are needed to foster the field of materiomics.

Unfortunately, in materiomics, tissue culture plates are seldom the substrate of choice; often we have to cope with small areas that provide only small amounts of material to analyse, be it cells, proteins or nucleic acids. These limitations pose a great challenge and condition the choice of assays.

In a seminal paper, Dalby and co-workers showed that particular topographies could influence the fate of human mesenchymal stromal cells (hMSCs). Standard immunohistochemical staining techniques were used to show how topographical cues could control cell fate (13). Based on these promising results, Unadkat *et al.* embarked on the development of a new platform that can generate thousands of different topographies and analyse their influence on cell fate as well as in other types of assays such as protein adsorption or even inhibition of bacterial growth (Figure 5.4) (17).

Whereas both authors relied on antibody-based assays, Unadkat *et al.* had to develop custom-made cell seeding devices and a new image analysis pipeline to acquire the resulting images automatically, certainly an interesting incentive to drive the field forward but hardly making it advance faster. Given the high number of features to be

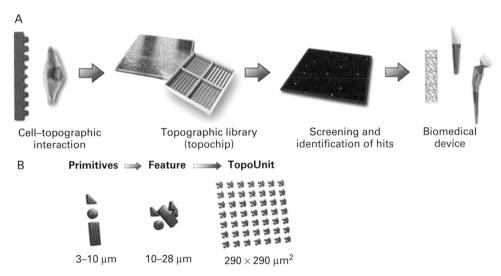

A

Cell–topographic interaction Topographic library (topochip) Screening and identification of hits Biomedical device

B **Primitives** ⇒ **Feature** ⇒ **TopoUnit**

3–10 μm 10–28 μm $290 \times 290\ \mu m^2$

Figure 5.4 A, schematic representation of a pipeline highlighting the steps needed from the assay development up to the implementation on a medical device. B, Using three primary shapes and combinations thereof, new features can be created and imprinted in polymers in order to test how they influence cell fate. Adapted with permission from (17).

analysed, assay reproducibility became paramount to identify the desired topographies. A relatively large number of replicas are needed in order to draw conclusions. As mentioned above, cell seeding and/or handling of the platform can affect assay reproducibility (some of the developed platforms have relatively poor adhesion capacity if not treated with cell adhesion proteins). Additionally, data mining in such experiments became critical (this topic will be discussed elsewhere in the book).

Besides topographical cues, Anderson *et al.* identified substrate chemistry as a key parameter controlling cell fate. The platform created comprised thousands of combinations of different monomers on a glass slide. Besides discovering combinations that control cell fate (using antibody-based assays), Anderson and co-workers also discovered that some surfaces could differentially bind proteins and so could control cell fate (12, 24, 25). Because cells never see bare materials but proteins previously adsorbed to the material, there is the interesting possibility that by providing a certain cocktail of proteins (such as the one present in serum in the cell culture medium) there will be differential adhesion of certain proteins to the detriment of others, depending on the physico-chemical characteristics of the substrate. Several technologies allow us to address this hypothesis, and such assays can be readily implemented in complex systems such as the ones mentioned above. Candidate proteins can be studied by tagging with fluorescent or radioactive labels, and their adherence to the surface of the materials can be analysed using high-content imaging technologies (discussed elsewhere) (26). Depending on the technology used for labelling, we can obtain qualitative or quantitative information about total protein adhesion, as well as spatial information which can be of utmost importance in platforms where the shape of the substrate varies. When no candidate protein is available, technologies such as mass and Raman spectrometry can be used to identify all the known proteins in a qualitative and/or quantitative manner (26–30). This highlights the fact that assays other than cell-based can be used to predict cell fate and ultimately simplify the type of assays that can be performed in complex platforms.

Box 5.2 Classic experiment

To identify new therapeutic candidates, two major approaches can be used in designing an assay for the initial discovery process: phenotypic-based screen or target-based screen. Interestingly, the recent major efforts of the larger pharmaceutical companies have been into target-based approaches. As the numbers suggest, careful consideration of the phenotypic end-point, how this relates to the pathology of the disease being targeted in combination with potential target candidates, will give one the highest chance of success in making a medical impact. This type of approach is highlighted in a recent publication in *Science* (3) and a proof of concept study in specifically targeting stem cells in articular cartilage.

For decades, approaches to cartilage repair have focused on surgical replacement of artificial matrices for *ex vivo* expanded chondrocytes or mesenchymal stem cells (MSCs) into the defective joint, marrow stimulation techniques or simple pain alleviation. With each of these approaches, the cartilage has yet to be restored to the native hyaline structure and thus the patient can be left with pain or loss of mobility.

Recently, Johnson *et al.* described a HTS-approach to identifying molecules which could direct the differentiation of bone-marrow-derived MSCs towards chondrocytes (3). Chondrogenic nodules, an early indicator of chondrogenesis, were positively stained with a membrane lipid binding dye, to identify the early cell condensation phenotype associated with the induction of chondrogenesis. Positive inducers of nodule formation were identified first by high-content imaging and verified for specificity by the presence of type II collagen, Sox9, aggrecan expression and the lack of osteoblastic phenotypic modulators. Key to the success of this study, all lead candidates were counterscreened for another desirable phenotype – inhibition of matrix degradation. The molecule termed Kartogenin (KGN) was identified in this way as having both protective and regenerative properties (Figure 5.5), and its efficacy was demonstrated in two mouse models of cartilage damage. Subsequently, an interaction between KGN and filamin A (FLNA) was demonstrated. FLNA typically sequesters CBF in the cytosol. In the presence of KGN, FLNA was no longer bound to CBF; thus it translocated to the nucleus where it activated Runx1-mediated transcription of early chondrogenic genes. Overall this paper high-lights a phenotypic approach to drug discovery that revealed a novel mechanism of action and will lead to future explorations in this arena for cartilage repair.

When using antibody-based assays (most of the assays in the field of materiomics fall in this category), a candidate protein and/or genes known to play a role in the chosen cellular pathway needs to be selected upfront. However, in many cases, information on potential targets is unavailable or is difficult to infer. Recently, Treiser and co-workers have shown that certain cell-shape parameters could have predictive power regarding the capacity of hMSCs to differentiate into the osteogenic lineage (31–33). In an elegant experiment, the authors showed that as early as 6 h after cell seeding one could, solely from cell shape, predict whether a MSC would become an osteoblast. Although in this case Treiser *et al.* used an end-point assay (phalloidin staining), there are possibilities of measuring cell-shape parameters in living cells. For example, Lifeact, a small 17-amino acid peptide derived from a yeast protein that can fluorescently label actin filaments without interfering with the filament dynamics, can be used for real-time monitoring of cell-shape-related events (34). The advantage of this technology is obvious: we can now use a few cells and follow the outcome individually using real-time imaging without the need for several replicated experiments. The capacity to predict bone-forming potential in 6 h is a huge improvement over other technologies such as gene and protein expression analysis. However, materials with pronounced changes in topography or with 'strong' topographies can limit the application of this approach. External signals perceived by the cells converge intracellularly to a distinct outcome such as phosphorylation of certain proteins, production of certain secondary messengers and recruitment of certain adapter proteins. The activation or modulation of these cellular networks directs the phenotypical

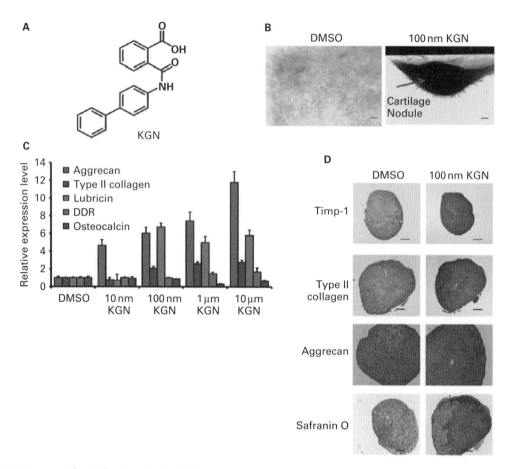

Figure 5.5　After testing thousands of different compounds for their capacity to induce nodule formation, one compound was selected (A and B); confirmation assays showed that the compound was able to induce a chondrogenic phenotype at different concentrations as shown by the upregulation of chondrogenic markers such as aggrecan and type II collagen (C and D). Adapted with permission from (3).

consequences of these signals. Biological understanding of these effects will allow us to design assays specifically aiming at those targets in these networks, opening the possibility to design specific assays and reducing assay complexity. At the moment, several technologies compatible with complex platforms can be used to analyse how cells respond to certain stimuli. Materials influence cells at various levels leading to changes in their gene expression profile and consequently in the secreted proteins, therefore posing pertinent biological questions.

The development of novel, more sensitive and quantitative, assays to measure gene expression has improved the analysis of the genome and measurement of gene expression. At this moment technologies exist that allow measurement of the expression of large numbers of genes in individual cells (35, 36). The basis of many of these technologies is the polymerase chain reaction (PCR), which has greatly facilitated the analysis of the genome as well as gene expression *in vitro* as well as in *in vivo* systems for many years. The power of this technique allows us to amplify and quantify signals from very small numbers of cells or amounts of tissue. Use of these technologies in materiomics will solve one of the major limitations in the field, the low number of cells, and broaden the analysis from a few targets (common in antibody-based assays or even reporter systems) to several, since using the same cell lysate one can analyse the expression of several target genes.

The use of reporter systems containing the promoter of genes of interest coupled to fluorescence or luminescent reporters will allow us to monitor cell fate without the use of antibodies and therefore will reduce the handling and some variation in the procedure. Several reporters are available to monitor the regulation of different pathways leading to cell proliferation, differentiation or even apoptosis (37–39). Moreover, specific parts of a signalling pathway can be analysed using reporter systems and more than one reporter can be used simultaneously, therefore increasing the throughput.

Fluorescence is a widely used optical method for biosensing owing to its selectivity and sensitivity. Development and application of fluorescence-based tools allow the monitoring and visualization of living cells (37–40). Detailed understanding of biological events – also as response to materials – is increasingly dependent on the ability to visualize and quantify signalling molecules and structural proteins as well as biomarkers and changes in those with high spatial and temporal resolution. Cells can undergo physiological changes such as activation processes or pathological stresses, and this response can be quite heterogeneous in time and space. Conventional biochemical analytical methods often fail to monitor such processes in bulk (such as in tissues or cellular suspensions) and often fail to resolve small differences in the kinetics and order of changes between different cells. Fluorescent molecules are valuable tools for such studies – they can be used as biomolecular labels, enzyme substrates, environmental indicators and cellular stains (41, 42). With these possibilities, they represent indispensable tools for materiomics research.

Similar to fluorescence, and probably more relevant to the field owing to the absence of background signals, is luminescence. Whereas some materials fluoresce at certain wavelengths, no material will emit luminescence. Therefore, luminescent technologies have excellent signal-to-noise ratios. Luminescent assays are based on proteins isolated from organisms that emit light as a communication tool, such as fireflies or certain jellyfish. These proteins are commonly used in reporter assays that monitor activation of signal transduction pathways. Moreover, spatial information can be obtained by using luminescence in combination with CCD (charge-coupled device) cameras, although the technology is still relatively expensive for use in academic laboratories (37).

Energy-transfer-based assays such as FRET (Förster or fluorescence resonance energy transfer) assays have recently become feasible in biomedical research (43–45). FRET generates fluorescence signals, which are sensitive to molecular conformation, association and separation in the 1 to 10 nm range. Hence, FRET is capable of resolving molecular interactions and conformations with higher spatial resolution than conventional optical microscopy. The principle of FRET is based on the use of a donor fluorophore in an excited electronic state, which may transfer its excitation energy to a nearby acceptor fluorophore. This resonance energy transfer can yield a significant amount of information on the selected donor–acceptor pair molecules. The strong distance dependence of FRET efficiency has been widely used for studying the structure and dynamics of proteins and nucleic acids, for the detection and visualization of inter- and intramolecular association and conformational changes, and for measuring changes and interactions between molecules.

Continuing advances in studying gene and protein function have triggered the need for high-sensitivity and high-affinity molecular biological probes for qualitative and quantitative detection (molecular beacons). Increasing evidence indicates that RNA molecules have a wide range of functions in living cells, ranging from conveying and interpreting genetic information through their role as intermediate molecules, to providing structural support for molecular machines, to gene silencing (46, 47). These functions are represented by their expression level, turnover rates and spatial distribution within cells. The possibility of imaging specific RNAs in living cells in real time allows information to be obtained on RNA synthesis, gene expression level, RNA processing, transport and localization. This promotes new opportunities for progress in materiomics and the optimization of biomaterials.

To obtain a mechanistic understanding of the cellular pathways and networks leading to a specific cellular response and/or phenotypes, one needs to identify and study the role of individual proteins in these responses. The field of functional genomics uses genomic tools such as cDNA over-expression and RNA interference (RNAi) technologies to manipulate the expression of individual genes in cells. Full genome sequencing and subsequent annotation of several genomes, including human, has facilitated the development of genome-wide collections of these reagents. In particular, genome-wide collections of synthetic short interfering RNAs (siRNAs) or short hairpin RNAs (shRNAs) have been generated and used to knock-down gene expression in combination with cell-based readouts (48, 49). This approach facilitates the discovery of individual genes, signalling pathways or protein networks involved in the biological response measured. Because several different technologies can be used to introduce these reagents into cells including transfection, electroporation and infection, a large range of cell types, extending to primary cells and even *in vivo*, can be used. In addition, integration of virally encoded shRNAs can generate stable knock-down of gene expression allowing for the study of cellular phenotypes developing over a longer time period (50, 51). Similar to small molecule screening, RNAi screens have been adapted to multi-well formats including 384 well plates. As a consequence, this technology can also be used in combination with high-content/high-throughput microscopy. A further development is the use of cell arrays, a technology where siRNAs are spotted on glass slides in a transfection mixture. When cells are cultured on these slides, they take up the siRNAs, causing knock-down of gene expression in these cells. The consequence of knock-down can be monitored in living cells or after fixation and staining with specific dyes or antibodies. In the field of materiomics, one can combine these different technologies, using different materials pre-coated with specific siRNAs to study the consequences of knock-down of specific genes under a specific condition. The further development of the technologies will allow high-throughput and high-content readouts for these systems.

5.3 Future perspectives

The development of bioassays in laboratories and clinics around the world continues to increase at a rapid pace. As the understanding of complex cell–material interactions and outside–in signalling deepens, high-throughput systems that can also incorporate physiologic microenvironments, matrix and paracrine signalling, and multimodal external actuation (mechanical, electrical, chemosensory) will increasingly become the mainstay of biological assays. Assays that can bypass the need for full-fledged, multilayered '-omics' analysis and instead substitute this with systems biology information integrated with phenomenological methods are not yet available but are very much needed. Once developed and validated, such assays will be efficient at cell fate classification and prediction of cell reprogramming outcomes. Understanding how the innate genetic code is modified and navigated by external stimuli can drive newer, clinically translatable biological outcomes, drugs and devices. To this end, multimodal assays that can combine gene-level readouts with structural information about cell states and protein and lipid compositions may be feasible in the future, for example relying on broad-band coherent anti-Stokes Raman spectroscopy and confocal imaging or FRET spectroscopy. The integration of current high-throughput platforms with the latest imaging-based assays holds enormous potential for increasing efficiency in the discovery of environmental influences on cellular behaviour. As the palette of fluorescent probes has grown enormously over the past decade, there have been numerous applications for these molecules in the visualization of dynamic protein structures, RNA and DNA (52, 53). In addition, spectroscopy-based methods have emerged in the comprehensive characterization of cellular components, including mass spectroscopy methods for the analysis of post-translational modifications of histones (54).

For a number of years the limitations of 2D cell culture and the relevance of 3D cell systems have become increasingly evident. Most assays described so far are carried out in tissue culture plates and therefore use cells that see and feel two dimensions. However, with the advent of rapid prototyping technologies, new materials and the use of scaffolds to grow and deliver cells, new challenges need to be addressed (Figure 5.6). Assays need

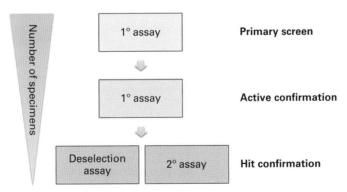

Figure 5.6 Schematic representation of a pipeline suitable for a materiomics-type assay. Note that in contrast with the previously described pipeline there are fewer steps involved, mostly owing to the nature and availability of the platforms to be tested.

to be performed in those 3D constructs to preserve spatial information and the first steps are under way. Some of the above-mentioned technologies will certainly be implemented in 3D assays in the coming years.

The possibility to monitor online some of the above-mentioned parameters (for example actin fibres, intracellular production of metabolites and secondary messengers, gene expression, cellular differentiation) will create new possibilities and technological challenges during the development of assays, but will certainly disclose new interactions not yet known.

5.4   Snapshot summary

- The increase in the number of synthesized compounds led to the development of new technologies.
- Mimicking the cellular environment is paramount during screening assays.
- Defining the biological target will dictate the choice of assay.
- The more confirmation assays available, the smaller the number of false positives.
- Online monitoring should be implemented in preference to end-point assays.
- Combining different technologies will contribute to the discovery process.
- Controls are essential to overcome intrinsic biological variations.
- Third dimensionality is critical for clinical translation.

Further reading

Mei Y, Saha K, Bogatyrey SR *et al*. Combinatorial development of biomaterials for clonal growth of human pluripotent stem cells. Nat Mater. 2010; **9**:768–78.

Rigbolt KT, Prokhorova TA, Akimov V *et al*. System-wide temporal characterization of the proteome and phosphoproteome of human embryonic stem cell differentiation. Sci Signal. 2011;**4**, rs3.

Rugg-Gunn PJ, Cox BJ, Lanner F *et al*. Cell-surface proteomics identifies lineage-specific markers of embryo-derived stem cells. Devel Cell. 2012;1–15.

Lee MJ, Ye AS, Gardino AK, Heijink AM *et al*. Sequential application of anticancer drugs enhances cell death by rewiring apoptotic signaling networks. Cell. 2012 May 11;**149**(4):780–94.

Sachlos E, Risueño RM, Laronde S *et al*. Identification of drugs including a dopamine receptor antagonist that selectively target cancer stem cells. Cell. 2012 Jun. 8;**149**(6):1284–97.

Kshitiz, Hubbi ME, Ahn EH *et al*. Matrix rigidity controls endothelial differentiation and morphogenesis of cardiac precursors. Sci Signal. 2012;**5**(227):ra41–1.

Trappmann B, Gautrot JE, Connelly JT *et al*. Extracellular-matrix tethering regulates stem-cell fate. Nat Mater. 2012;**11**(7):1–8.

References

1. Kohler G, Milstein C. Continuous cultures of fused cells secreting antibody of predefined specificity. Nature. 1975;**256**(5517):495–7.
2. Riechmann L, Clark M, Waldmann H, Winter G. Reshaping human antibodies for therapy. Nature. 1988;**332**(6162):323–7.

3. Johnson K, Zhu S, Tremblay MS, Payette JN *et al*. A stem cell-based approach to cartilage repair. Science. 2012;**336**(6082):717–21.

4. Flaim CJ, Chien S, Bhatia SN. An extracellular matrix microarray for probing cellular differentiation. Nat Methods. 2005;**2**(2):119–25.

5. Yliperttula M, Chung BG, Navaladi A, Manbachi A, Urtti A. High-throughput screening of cell responses to biomaterials. Eur J Pharmaceut Sci. 2008;**35**(3):151–60.

6. Miraoui H, Marie PJ. Fibroblast growth factor receptor signaling crosstalk in skeletogenesis. Sci Signal. 2010;**3**(146):re9-re.

7. Siddappa R, Fernandes H, Liu J, van Blitterswijk C, de Boer J. The response of human mesenchymal stem cells to osteogenic signals and its impact on bone tissue engineering. Curr Stem Cell Res Therapy. 2007;**2**(3):209–20. Epub 2008/01/29.

8. Macarron R, Banks MN, Bojanic D *et al*. Impact of high-throughput screening in biomedical research. Nat Rev Drug Discovery. 2011;**10**(3):188–95.

9. Wan C, Gilbert SR, Wang Y *et al*. Activation of the hypoxia-inducible factor-1alpha pathway accelerates bone regeneration. Proc Natl Acad Sci. 2008;**105**(2):686–91.

10. Hubbell J. Synthetic biomaterials as instructive extracellular microenvironments for morphogenesis in tissue engineering. Nat Biotechnol. 2005;**23**:47–55.

11. Langer R, Vacanti JP. Tissue engineering. Science. 1993;**260**(5110):920–6.

12. Anderson DG, Levenberg S, Langer R. Nanolitre-scale synthesis of arrayed biomaterials and application to human embryonic stem cells. Nat Biotechnol. 2004;**22**(7):863–6.

13. Dalby MJ, Gadegaard N, Tare R *et al*. The control of human mesenchymal cell differentiation using nanoscale symmetry and disorder. Nat Mater. 2007;**6**(12):997–1003.

14. Khan F, Tare RS, Kanczler JM, Oreffo ROC, Bradley M. Strategies for cell manipulation and skeletal tissue engineering using high-throughput polymer blend formulation and microarray techniques. Biomaterials. 2010;**31**(8):2216–28.

15. Lovmand J, Justesen J, Foss M *et al*. The use of combinatorial topographical libraries for the screening of enhanced osteogenic expression and mineralization. Biomaterials. 2009;**30**(11):2015–22.

16. McMurray RJ, Gadegaard N, Tsimbouri PM *et al*. Nanoscale surfaces for the long-term maintenance of mesenchymal stem cell phenotype and multipotency. Nat Mater. 2011;**10**(8):637–44.

17. Unadkat HV, Hulsman M, Cornelissen K *et al*. An algorithm-based topographical biomaterials library to instruct cell fate. Proc Natl Acad Sci. 2011;**108**(40):16565–70.

18. Caplan AI. Adult mesenchymal stem cells for tissue engineering versus regenerative medicine. J Cell Physiol. 2007;**213**(2):341–7.

19. Doorn J, Moll G, Le Blanc K, van Blitterswijk C, de Boer J. Therapeutic applications of mesenchymal stromal cells; paracrine effects and potential improvements. Tissue Eng B: Rev. 2011.

20. Birmingham A, Selfors LM, Forster T *et al*. Statistical methods for analysis of high-throughput RNA interference screens. Nat Methods. 2009;**6**(8):569–75. Epub 2009/08/01.

21. Clark NR, Ma'ayan A. Introduction to statistical methods for analyzing large data sets: gene-set enrichment analysis. Sci Signal. 2011;**4**(190):tr4-tr.

22. Malo N, Hanley JA, Cerquozzi S, Pelletier J, Nadon R. Statistical practice in high-throughput screening data analysis. Nat Biotechnol. 2006;**24**(2):167–75.

23. Vaux DL, Fidler F, Cumming G. Replicates and repeats – what is the difference and is it significant? EMBO Rep. 2012;**13**(4):291–6.

24. Mei Y, Saha K, Bogatyrev SR *et al*. Combinatorial development of biomaterials for clonal growth of human pluripotent stem cells. Nat Mater. 2010;**9**(9):768–78.

25. Saha K, Mei Y, Reisterer CM *et al.* Surface-engineered substrates for improved human pluripotent stem cell culture under fully defined conditions. Proc Natl Acad Sci USA. 2011;**108**(46):18714–19.

26. Barradas AMC, Lachmann K, Hlawacek G *et al.* Surface modifications by gas plasma control osteogenic differentiation of MC3T3-E1 cells. Acta Biomater. 2012:**8**:1–9.

27. Rigbolt KT, Prokhorova TA, Akimov V *et al.* System-wide temporal characterization of the proteome and phosphoproteome of human embryonic stem cell differentiation. Sci Signal. 2011;**4**(164):rs3. Epub 2011/03/17.

28. Rugg-Gunn PJ, Cox BJ, Lanner F *et al.* Cell-surface proteomics identifies lineage-specific markers of embryo-derived stem cells. Devel Cell. 2012:1–15.

29. Schirle M, Bantscheff M, Kuster B. Mass spectrometry-based proteomics in preclinical drug discovery. Chem Biol. 2012;**19**(1):72–84.

30. Steinhauser ML, Bailey AP, Senyo SE *et al.* Multi-isotope imaging mass spectrometry quantifies stem cell division and metabolism. Nature. 2012:**481**;1–5.

31. Kumar G, Tison CK, Chatterjee K *et al.* The determination of stem cell fate by 3D scaffold structures through the control of cell shape. Biomaterials. 2011;**32**(35):9188–96.

32. Liu E, Treiser MD, Patel H *et al.* High-content profiling of cell responsiveness to graded substrates based on combinatorially variant polymers. Comb Chem High Throughput Screen. 2009;**12**(7):646–55. Epub 2009/06/18.

33. Treiser MD, Yang EH, Gordonov S *et al.* Cytoskeleton-based forecasting of stem cell lineage fates. Proc Natl Acad Sci. 2010;**107**(2):610–5.

34. Riedl J, Crevenna AH, Kessenbrock K *et al.* Lifeact: a versatile marker to visualize F-actin. Nat Methods. 2008;**5**(7):605–7.

35. Longo D, Hasty J. Dynamics of single-cell gene expression. Mol Syst Biol. 2006;**22**: doi:10.1038/msb4100110.

36. White AK, VanInsberghe M, Petriv OI, Hamidi M, Sikorski D, Marra MA, *et al.* High-throughput microfluidic single-cell RT-qPCR. Proc Natl Acad Sci. 2011;**108**(34):13999–4004.

37. de Boer J, van Blitterswijk C, Löwik C. Bioluminescent imaging: Emerging technology for non-invasive imaging of bone tissue engineering. Biomaterials. 2006;**27**(9):1851–8.

38. Dower SK, Qwarnstrom EE, Kiss-Toth E. Fluorescent protein reporter systems for single-cell measurements. Methods Mol Biol. 2007;**411**:111–9.

39. Liu J, Barradas A, Fernandes H, Janssen F *et al.* In vitro and in vivo bioluminescent imaging of hypoxia in tissue-engineered grafts. Tissue Eng C Methods. 2010;**16**(3):479–85. Epub 2009/08/19.

40. Conrad C, Wunsche A, Tan TH *et al.* Micropilot: automation of fluorescence microscopy-based imaging for systems biology. Nat Methods. 2011;**8**(3):246–9.

41. Hirsch JD, Haugland RP. Conjugation of antibodies to biotin. Methods Mol Biol. 2005;**295**:135–54.

42. Steinberg TH, Agnew BJ, Gee KR *et al.* Global quantitative phosphoprotein analysis using Multiplexed Proteomics technology. Proteomics. 2003;**3**(7):1128–44.

43. Demarco IA, Periasamy A, Booker CF, Day RN. Monitoring dynamic protein interactions with photoquenching FRET. Nat Methods. 2006;**3**(7):519–24.

44. Sun Y, Day RN, Periasamy A. Investigating protein–protein interactions in living cells using fluorescence lifetime imaging microscopy. Nat Protocols. 2011;**6**(9):1324–40.

45. Sun Y, Wallrabe H, Booker CF, Day RN, Periasamy A. Three-Color spectral FRET microscopy localizes three interacting proteins in living cells. Biophys J. 2010;**99**(4):1274–83.

46. Mendell JT, Olson EN. MicroRNAs in stress signaling and human disease. Cell. 2012;**148**(6):1172–87.

47. Pasquinelli AE. MicroRNAs and their targets: recognition, regulation and an emerging reciprocal relationship. Nature Rev Genet. 2012;**13**(4):271–82.

48. Orvedahl A, Sumpter R Jr, Xiao G *et al*. Image-based genome-wide siRNA screen identifies selective autophagy factors. Nature. 2011;**480**(7375):113–7.

49. Paul P, van den Hoorn T, Jongsma MLM *et al*. A genome-wide multidimensional RNAi screen reveals pathways controlling MHC Class II antigen presentation. Cell. 2011;**145**(2):268–83.

50. Bernards R, Brummelkamp TR, Beijersbergen RL. shRNA libraries and their use in cancer genetics. Nat Methods. 2006;**3**(9):701–6.

51. Mullenders J, Fabius AW, Madiredjo M, Bernards R, Beijersbergen RL. A large scale shRNA barcode screen identifies the circadian clock component ARNTL as putative regulator of the p53 tumor suppressor pathway. PLoS ONE. 2009;**4**(3):e4798. Epub 2009/03/12.

52. Jain A, Liu R, Ramani B *et al*. Probing cellular protein complexes using single-molecule pull-down. Nature. 2011;**473**(7348):484–8.

53. Schwalb NK, Temps F. Base sequence and higher-order structure induce the complex excited-state dynamics in DNA. Science. 2008;**322**(5899):243–5.

54. Downes A, Mouras R, Bagnaninchi P, Elfick A. Raman spectroscopy and CARS microscopy of stem cells and their derivatives. J Raman Spectrosc. 2011;**42**(10):1864–70.

6 High-content imaging

Frits Hulshof, Er Liu, Andrea Negro, Samy Gobaa, Matthias Lutolf,
Prabhas V. Moghe and Hugo Fernandes

Scope

High-content imaging (HCI) plays a pivotal role in high-throughput screening (HTS) of biological responses to biomaterials. This chapter will give a brief introduction to its basic principles by explaining the most commonly used imaging techniques, describing the general structure of an image analysis pipeline and providing a summary of available HCI software. Special emphasis will be devoted to the initial steps of image analysis such as image correction, segmentation and feature extraction. Additionally we include a brief overview of relevant literature and description of promising new tools for HCI. Finally, two classic experiments will be described, which use state-of-the-art HCI in biomaterial science by experts in the field.

6.1 Origins of high-content imaging

The first person to observe micro-organisms with a microscopic device was the Dutchman Antoni van Leeuwenhoek (1), who was able to make and polish tiny lenses of high curvature which were the forerunners of the modern microscope. Robert Hooke, the English father of microscopy, later improved on the design and confirmed van Leeuwenhoek's findings. Since then, light microscopy has evolved into its modern incarnation. The first use of fluorescence microscopy in biology was in 1881 when the bacteriologist Paul Ehrlich used fluorescin to observe the aqueous humour in the eye. The first immuno-staining was performed in 1950 by Melvin Kaplan (2). Green fluorescent protein (GFP) was the first fluorescent protein used in biology. It was isolated from jellyfish in 1961 by Shimomura and co-workers, and it was cloned by Prisher in 1992. The ability either to use antibody conjugated fluorophores or to clone GFP into chimaeric proteins led to the implementation of the fluorescence microscope in modern biology. High-content imaging (HCI) is the automation of fluorescence microscopy whereby the manual interpretation of images is replaced by computer algorithms.

Materiomics: High-Throughput Screening of Biomaterial Properties, ed. Jan de Boer and Clemens van Blitterswijk. Published by Cambridge University Press. © Cambridge University Press 2013.

The origins of HCI can be traced to the drug screening industry where it was developed more than 10 years ago. The possibility of automatically acquiring images from thousands of individual wells, which contain various compounds, became paramount to the pharmaceutical industry. Prior to automation, one of the major hurdles was the analyses and interpretation of images by human experts, introducing unwanted variations in addition to the time required to analyse large datasets. The drug screening industry realized that HCI had advantages over other more commonly used high-throughput techniques, such as fluorescence assisted cell sorting (FACS), gene expression profiling with microarrays or next-generation sequencing. Those techniques often lack the ability to resolve the spatial and temporal features (3). HCI not only provides the necessary resolution but also enables the tracking of organelles or even specific receptors. The academic world quickly caught up with the pharmaceutical industry in the development of HCI (4).

One of the great advantages of imaging is the ability to analyse single cells instead of averages of cell populations. This enhances the measurement of biological events, since a cell population is never homogeneous, and taking the average data of a population masks interesting phenotypes of individual cells. The data generated by HCI is more consistent, the output can be standardized, and it can be stored in online databases, where it can be easily retrieved and interrogated (5). A disadvantage of imaging is that it can be difficult to obtain strong statistical power in discriminating a dataset. To overcome this, it is vital to achieve clear and distinct readouts with strong controls. This part of assay-development was covered in Chapter 5.

Another disadvantage is the sheer size of the datasets resulting from large-scale HCI experiments. Not only is the raw dataset often daunting but the amount of data is also multiplied by every correction step. The average computer server is often unable to handle the data within a realistic timeframe, creating the need for a cluster computer setup and effective data logistics.

6.2   High-content imaging techniques

Most of the conventional HCI systems are based on fluorescence microscopy. The basic principle of these image-based assays is very simple: intensity quantification and relative locations of fluorophores in the specimen. This means that it is possible to visualize many different biological molecules, since fluorophores can be conjugated to many different substrates such as proteins, nucleotides and lipids. Alternatively, fluorescent molecules can be indirectly bound to targets by using labelled antibodies or other peptides with specific binding properties. Figure 6.1 shows examples of fluorescent markers used in HCI. Alternatively, reporter cell lines can be used.

Currently the most common instrument for fluorescence microscopy is an inverted fluorescence wide-field microscope such as the Zeiss Axiovert series, Olympus IX series, Leica DM series or Nikon eclipse Ti series. For high-throughput screens such microscopes need to be fully automated. A fully motorized stage and a filter cube turret or filter wheels are essential.

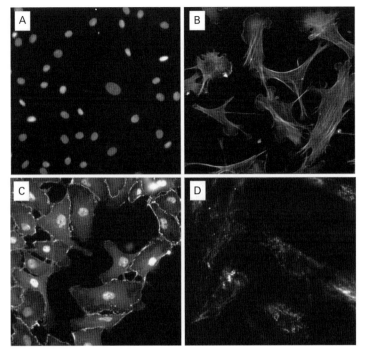

Figure 6.1 Images of various cells stained with commonly used markers taken with BD Pathway 435 automated fluorescence microscope. A, 4′,6-diamidino-2-phenylindole (DAPI) marks the nuclei of cells by binding to DNA. It is commonly used to quantify cells or as a nuclear counterstain. B, Phalloidin is a peptide toxin that specifically binds filamentous actin. Phalloidin, coupled to a fluorophore, is a very useful tool to visualize actin stress fibres and lamellipodia, and it is also used to observe cell shape. C, Zona Occludens 1 (ZO1) is a protein that is part of tight-junction complexes. Immuno-staining ZO1 is used to visualize tight junctions and the interconnectivity of cells growing in a monolayer. D, Tissue non-specific alkaline phosphatase (ALP) is a protein that is often used as an early marker for osteogenic differentiation.

Alternatively, when an assay is designed on a multi-well plate platform, an automated microscope such as the BD pathway or Opera can be advantageous, since these are optimized for multi-well plates and can also be fitted with live cell imaging options.

Another instrument suitable for HCI is the confocal microscope. Confocal microscopy can offer higher image quality than wide-field systems. Since the confocal system only detects light from a single z plane, it is possible to make image stacks of cells. Such image stacks can be deconvolved by specific algorithms which are able to determine the points from which the scattered light originates. This makes it possible to study subcellular features in high detail. The resolution in 'conventional' fluorescence microscopy techniques is limited by the diffraction limit of light. In effect, this means that objects that are smaller than 100–200 nm cannot be distinguished. Super-resolution fluorescence microscopy techniques are currently not suitable for high-throughput applications because of their long image acquisition times.

Most high-throughput experiments are done on fixed samples. Even though fixation introduces certain artefacts in the sample (6), it greatly simplifies the imaging process and makes it more robust. The option to use live-cell imaging provides new possibilities as well as new challenges. For example, it allows the study of changes in protein localization or dynamic processes such as receptor internalization. For live-cell imaging, the imaging conditions have to be optimal in order to not expose the cells or tissue to excessive light, which would cause photo-bleaching and photo-toxicity.

Imaging of cells and tissues on biomaterials brings additional challenges. Often material–cell interfaces do not form the necessary flat surface that is essential for high-quality fluorescence imaging. Uneven surfaces cause parts of the object to be out of focus and can also cause problems for the autofocus systems in automated setups. Often there is a need to study the material–cell interface in a 3D environment, which is prone to all of the above-discussed problems for uneven surfaces. It is therefore recommended to use a confocal system to capture the added z plane typical of a 3D system. *A priori* design of 3D systems with materials surfaces and structures suitable for HCI is critical for good image acquisition. As mentioned earlier a flat surface is preferable, and it also greatly simplifies automation if single units in an array have a uniform size and are spaced in a rational and consistent manner.

Box 6.1 Classic experiment

Combination of high-content and high-throughput fluorescence imaging for cell–biomaterial interactions

Tyrosine-derived polycarbonates, poly(desaminotyrosyl-tyrosine ethyl ester carbonate ((abbreviated as poly(DTE carbonate) and poly desaminotyrosyl-tyrosine octyl ester carbonate (poly(DTO carbonate)) homopolymers and their blends (70/30, 50/50, 30/70 ratio, by mass), were flow-coated on round glass cover slips. The coated slips exhibited well-characterized surface gradients in hydrophobicity and surface roughness (Figure 6.2A and (35)). This system offers a simple approach to realize combined variations of chemical composition and topographical features of the substrates within a single chip and study their effects on cell responses.

Human Saos-2 cells transfected with GFP-tagged farnesylation (GFP-f) gene were examined for their responsiveness to the surface texture gradients of the blends. The morphology and organization of Saos-2 cells were screened spatially across the polymer substrate chip in real time using the motorized stage of a confocal laser scanning microscope (CLSM) confined to a temperature-controlled chamber. Two levels of scanning were performed (Figure 6.2B,C): global screening was performed under 10× objective in automatic tile-scan mode for cell attachment and adhesion imaging, followed by an automated cell count using the ImagePro Plus software (including smoothing and thresholding with watershed algorithms to separate cells that share borders with each other). For HCI, high-resolution green fluorescent single-cell images were taken under a 63× objective for the quantitation of intracellular features of SAOS-2 cells expressing GFP-farnesylation (GFP-f) fusion proteins. Ten optical sections were taken on each cell at z-step size of 0.5 μm, and these image stacks then underwent average projection to generate a single image for subsequent image analysis.

Image-based morphometric feature extraction was performed on a series of image analysis pipelines through ImagePro Plus software (Media Cybenetics Inc.) (8). Multiple cells examined for each position on the roughness gradient were used to calculate a population distribution for each cell morphometric descriptor. The cell descriptors pool included: (i) cellular morphologic parameters; (ii) spatial distribution of farnesylated-protein and texture parameters; and (iii) protein farnesylation parameters. The comparison of cell descriptors along the roughness gradient was demonstrated in a 'heat map' representation of p-values from ANOVA test (Figure 6.2D).

Several new insights were gained about cell adhesive responses to dual composition–topography graded substrates. Cell adhesion was maximized at intermediate regions on the polymer chip, characterized by intermediate roughness and the steepest roughness gradient (39). Through HCI, different morphometric parameters of the organization and intensity of GFP-f were identified that correlate with the most adhesive substrate compositions (chemistry) or with the degree of surface roughness.

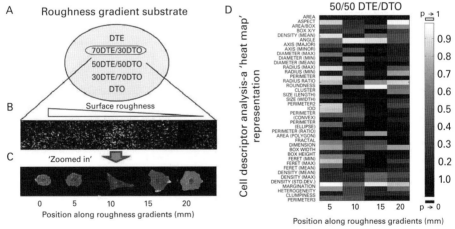

Figure 6.2 High-content/high-throughput screening of cells on combinatorially designed biomaterial chip. A, Overview of the design of the roughness gradient substrates based on polymer blends. Along the horizontal axis is a continuous temperature annealing gradient; along the vertical axis is the compositional variation of poly(DTE carbonate)/poly(DTO carbonate) blends. B, C, Example of high-throughput/high-resolution imaging on GFP-farnesylation Saos2 cells on 50/50 roughness gradient on the combinatorial chip at low magnification (10×) (B) and high magnification (63×) (C). D, Heat-map representation of a pool of cell descriptors derived from GFP-farnesylation Saos2 cells along roughness gradients. Reproduced with permission from (39).

6.3   The image analysis pipeline

When the 'raw' images have been acquired by the microscope, image analysis takes place. Typically this data goes through a so-called image analysis 'pipeline' which starts with the raw data and ends with a dataset or even a first overview of the results. Between the beginning and the end of the pipeline, there are several steps. A typical pipeline contains elements that perform the tasks shown in Figure 6.3. These tasks or steps are described in this chapter except for the data analysis (data mining), which is covered in the next chapter, because it relies heavily on mathematics, bioinformatics and advanced statistics.

The first step in the pipeline is generally an image correction or image pre-processing step. In this step the dataset is corrected for systematic errors caused by the imaging equipment, and errors caused by the limitations of the assay and biological variance. Flat-field correction is often required to compensate for uneven illumination of the sample caused by the spherical shape of the light bundle. This can be done by calculating the light curvature from a reference image and using the calculated curve to correct the dataset. Alternatively, algorithms that quantify the abnormal lighting from the dataset itself and correct for the uneven illumination can be used. Other systematic errors such as well-to-well variation and plate-to-plate variation have to be corrected. Well-to-well variation is often corrected by the median polish method which corrects the dataset

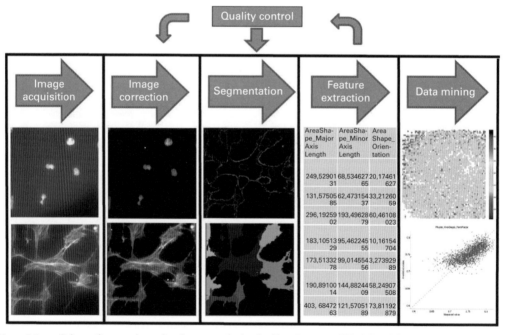

Figure 6.3   Schematic overview of an image analysis pipeline. Images of hMSCs with nuclear and filamentous actin stains. Segmentation images created with cell profiler.

stepwise by the median value in the data. In general, the higher the quality of the raw data, the less pre-processing required. The main function of the pre-processing stage is to increase the quality of the subsequent image segmentation and to make the segmentation more robust.

Once the images have been corrected, they are ready for segmentation. Segmentation is a process of partitioning an image into multiple non-intersecting regions (often called regions of interest or ROIs) such that each region is homogeneous and the union of no two adjacent regions is homogeneous (7). In this step, it is determined which aspects of the images will be quantified. Effectively the pixels within the image are divided into different populations on the basis of their intensity and (relative) positions. Simple examples of quantifiable features include cell number or number of nuclei for a proliferation assay, for which a nuclear probe such as 4′,6-diamidino-2-phenylindole (DAPI) is commonly used (Fig. 6.1A). Generally, in HCI, more complex features such as details in cell morphology, cell orientation and marker localization in phenotypic screens are studied. These complex details in cell morphology are sometimes also termed morphometric descriptors. Examples of these are average cell area, perimeter, and relative length of major and minor axes (8). Filamentous actin, directly labelled with an actin-GFP plasmid transfection or indirectly labelled with phalloidin (Figure 6.1B), for example, can be used to measure these cell morphological parameters. The expression of certain markers can be measured, such as alkaline phosphatase (Figure 6.1D), which is associated with bone differentiation of hMSCs, or Zona Occludens 1 (Figure 6.1C), which is part of functional tight junctions in epithelial cell layers. Correct image segmentation is very important as errors in this step lead to faulty metrics in the feature extraction step.

During the feature extraction step, the regions that have been identified in the segmentation step are measured and collected in datasets or databases. In this step, many metrics such as relative shapes of cells or subcellular features, intensities, intensity distributions and texture can be measured. Examples from the literature include: the determination of cell numbers on polymer surfaces (9–11), details in cytoskeletal morphology (8, 12), subcellular features (5, 13), cytoskeletal features (14), and autophagy (15).

Between different steps in the pipeline, it is important to implement some form of quality control to assess the quality of the assay, the segmentation and the dataset as a whole. The performance and the sensitivity of the assay have to be determined. There are many statistical methods that can be used (16). One of the most common methods to predict assay performance is the Z' factor, defined in Chapter 5. The Z' factor lets the researcher compare different assays and screens by using control wells (Z') and sample wells (Z).

To assess the sensitivity of an assay, most often the minimum significance ratio (MSR) is calculated. The MSR gives a measure for the reproducibility of potency values. When the quality of the assay is within the desired range it is ready for analysis. In the final step, the dataset can be analysed to identify hits, look for correlations between parameters and look for other trends and relationships. The data mining of large datasets requires powerful analysis software tools and advanced statistics. These are discussed in more detail in the next chapter (Chapter 7).

Box 6.2 Classic experiment

Quantifying stem cell fate in artificial niches

The fate of stem cells is largely determined by the microenvironment (niche) in which they reside(40). To gain more refined insights on stem cell/niche interactions, we have developed a novel cellular microwell array platform suitable for HCI to study the effects of substrate stiffness, tethered or soluble proteins and cell density (41).

To illustrate the custom-designed image analysis for the microwell array, an experiment is shown which intends to quantify osteogenic differentiation of hMSCs as a function of matrix stiffness and fibronectin concentration. Immunocytochemistry staining was performed to mark expression of the osteogenic marker ALP, and DAPI was used to label the nuclei. A Matlab program automatically segments the images in order to quantify the ALP signal. The analysis is based on two successive operations, namely the microwell detection in bright field images and osteogenic marker (ALP) segmentation in fluorescent images. The microwell detection starts with filtering steps such as background correction, median and Gaussian filtering. Then we convolve the array images with a template microwell. The resulting convolution maps show several peaks at the centre of each well. The microwell coordinates are then extracted and single well images cropped accordingly. To maintain the link between microenvironment composition and physical location on the array, we feed a Matlab database with single cropped images and their addresses within the arrays (Figure 6.4). To match the ALP signal to each nucleus, we use crops of single wells and perform adaptive thresholding: we take the 70% highest DAPI histogram values. The resulting ROI is used as a starting point from which we initiate the detection of the ALP signal in the differentiating cells.

In the above example, we implemented an imaging system able to quantify single-cell differentiation in microstructured hydrogel microwells. A consecutive multivariate analysis on the measured parameters demonstrated that osteogenic differentiation of hMSCs is controlled by the substrate stiffness but not by the concentration of adhesion ligands (41).

6.4   Software

In the years since HCI first made its way into academic research, many groups developing the technology have also developed their own image analysis software. Several groups have published the capabilities of their software and made them freely available to the scientific community. Examples of free available packages developed by academic groups include HCDC-HITS (17), Micropilot (18) (also requires Labview commercial software) and EBJImage (19).

In 'medium' high-throughput approaches, research groups often use different software packages for different stages of the image analysis pipeline. For example, image

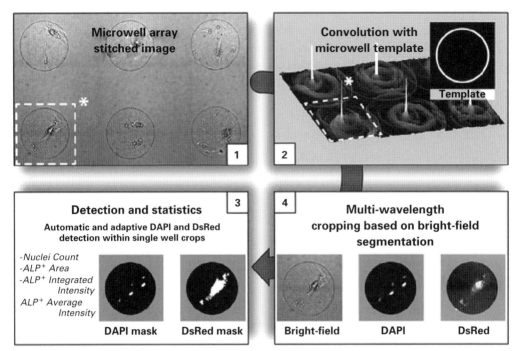

Figure 6.4   Pipeline of image acquisition and analysis for the quantification of osteogenic differentiation in hydrogel microwell arrays. 1, Bright-field reconstructed image of six contiguous microwells containing differentiating hMSCs. 2, Convolution map of the reconstructed mosaic of microwells. Peaks mark the physical centre of each microwell. 3, Crops of the microwells in three wavelengths. The cropping was performed based on the coordinates of every convolution peak as obtained in the previous step. 4, Quantification by measuring the DAPI and ALP signals in each microwell.

acquisition will use software from the microscope manufacturer, while image correction and data analysis are performed using other software. In contrast, when research groups regularly perform high-throughput experiments, it is more common to have the whole pipeline or workflow automated by the same software package to reduce manual work to a minimum.

Microscope manufacturers such as Leica, Zeiss and Olympus have also started providing comprehensive software packages for HCI with their microscope products. Often these software packages work well, but lack flexibility. Often the source-code is not open source, meaning that it is impossible to modify the software freely.

Examples of open-source software programs that are often used for certain elements of the image analysis pipeline are ImageJ and Cell Profiler. Many groups use shared or self-build plug-ins for ImageJ to handle many pre-processing, segmentation and measurement steps (15). ImageJ is a very flexible, lightweight, open-source platform for image handling and analysis (20). However, creating and editing scripts for Image J does require some simple knowledge of programming which the majority of biological scientists do not have. Fortunately, there is a large community of imaging experts who write and freely share their plug-ins for ImageJ to add functionality and help those scientists with insufficient experience in programming.

Cell Profiler is a program that was developed as a flexible tool to make image processing and analysis more accessible to those scientists without comprehensive programming experience (21). It allows the user to design image analysis pipelines by combining many different modules which are able to apply correction, segmentation and feature extraction steps on large image sets. Every module is customizable to improve flexibility, and new features are added in through regular updates.

6.5   High-content imaging in materiomics

In the past few years of HCI, the emphasis has shifted from increasing screening capacity (quantity) to increasing content and combining technologies (quality) (22). An example of combining technologies is the merge of high-throughput small RNA assays and HCI. Genome-wide libraries of small interfering RNAs (siRNA) enable the selective knock-down of expressed proteins. Multiparametric imaging has been used to associate specific protein deficiencies with cell phenotypes (12). To increase the specificity of HCI assays, FRET sensors (explained in Chapter 5) are increasingly being used, for example to detect live-cell caspase activation. Combining several caspase sensors enabled the observation of multiple decisive stages of cell death (23).

The trend of multiplexing HCI with other technologies is clearly present in the biomaterials field, allowing high-throughput biomaterials research or 'materiomics'. The field of materiomics comprises the study of structure–activity relations of an array of materials with physico-chemical and biological properties. As biomaterials have evolved from design of a small number of disparate materials to larger libraries of combinatorial engineered materials, HCI approaches can now be more widely deployed to dissect cell–biomaterial interactions. Several groups in the biomaterials field have proposed various methods for the use of HCI of cells, and a brief summary of these appears below. The methods chosen vary from straightforward methods to determine cell numbers attached to substrates, for example, to more complex methods to determine cell morphology and expression of molecular markers. For example, the Anderson group has published several articles on the applications of their polymer microarrays for HTS. Applications include the investigation of the clonal growth of embryonic stem cells (24), embryonic body adhesion (9) and Islet of Langerhans attachment (10).

Laser scanning cytometry (LSC) was used to study cell adhesion and expression of markers of certain cell types on various polymers (25). In essence, LSC is similar to flow cytometry in that it identifies individual cells before measurements. The main difference is that it works on cells on flat surfaces instead of in a capillary flow. As a technique it is suitable for (semi-)high-throughput quantitative imaging of flat surfaces, allowing rapid quantification of cell numbers and cellular fluorescence intensities. However, LSC lacks the resolution and optical sectioning of confocal microscopy necessary for ultra-structural analysis of subcellular features.

The Bradley group also used microarrays of spotted polymer blends to study cell–biomaterial interactions in high-throughput formats. Instead of LSC, an automated inverted fluorescence microscope is used to determine cell numbers on different polymer blends. Examples of the cellular parameters quantified in these studies include: the attachment of K562 human erythroleukaemic suspension cells (26), L929 mouse fibrosarcoma cells (11) and human skeletal stem cells on polymer blend spots (27).

In contrast to these approaches, biomaterial–cell interactions have been extracted by discerning complex cellular parameters such as various intensity, texture and geometric details of cellular and intracellular morphology and organization (8). This work is an effective example of quantitative measurements of complex cellular parameters in the context of biomaterial research. In this study, GFP reporter cell lines were used in lieu of analysing fixed cells labelled with antibody-based immuno-staining, and various GFP plasmids were used to stain different adhesion and cytoskeletal molecules in the cell such as actin, microtubules, plasma membrane and focal adhesions. Fixed cells were imaged by a motorized confocal microscope. Using image processing, the various cytoskeletal elements were segmented, measured and combined to create morphometric descriptors. These morphometric descriptors were able to describe many features of the cells, which could then be coupled to the substrate material properties.

In subsequent work similar, morphometric descriptors – based on the actin cytoskeleton – were used to forecast stem-cell lineage fates that were induced by various substrata and growth factors (14). High-throughput HCI has been used to study the effects of microtopographies on multiple cellular parameters such as marker expression (28) and cell morphology. Previous work concerning the biological effects of nanoscale (29) and microscale topographies (30) also used fluorescence microscopy, but not in a high-throughput, high-content fashion. For HTS, cells were cultured on a polymeric surface containing thousands of randomly generated topographies (the TopoChip) and imaged by fluorescence microscopy. The various topographies in $300 \times 300 \ \mu m^2$ 'microwells' (TopoUnits) were arrayed 66 by 66 on a $2 \times 2 cm^2$ chip. Because of the constant dimensions in the TopoChip design, the automated microscope (BD Pathway 435), which is optimized for well-plates, could be programmed to automatically image every TopoUnit in multiple channels. This could be achieved by creating a 'coordinate map' of the TopoUnits and perfectly aligning the TopoChip placed in the microscope to this map. The resulting images were analysed using a custom cell profiler pipeline. Analysis showed significant effects of the different microtopographies on cell morphology and ALP expression of hMSCs (28).

Finally, HCI has been used to study cell–biomaterial interactions on soft hydrogel microwell arrays. A motorized fluorescence microscope with live-cell imaging capabilities was used. This made it possible to obtain phase contrast images of the arrays at early time points, before fixing the cells for immuno-staining at final time points. Human mesenchymal stromal cells and mouse neural stem cells were stained for various markers to quantify differentiation into different lineages. The marker intensities were quantified and used as a measure of differentiation.

6.6   Future perspectives

The impact of material-based combinatorial libraries will be understood only with the advent of methods that can rapidly evaluate the impact on cell fate of those thousands of combinations. Currently, as the biological field enters the '-omics' era, many imaging techniques are being integrated in more high-throughput and/or high-content formats. Some of these techniques show promise for application in the biomaterials field. In this final part of the chapter, advances in two of these methods – FLIM FRET and Raman spectroscopy – will be briefly reviewed.

There are many ways to detect FRET (covered in Chapter 5); the two most common approaches are spectral radiometric imaging and fluorescence lifetime imaging (FLIM). (31) Spectral radiometric imaging detects FRET by measuring intensities; it is straightforward to implement instrumentally and has relatively short times. The major drawback is that the complete experimental setup has to be extensively calibrated, from the optical system itself to the actual sample. FLIM detects FRET by measuring the lifetime, rather than the intensity of the fluorescent signal, leading to a more robust measurement because it is usually independent of fluorophore concentration, excitation and detection efficiencies. However, FLIM has the drawbacks of relatively longer acquisition times and a requirement for more specialized instrumentation.

In spite of these shortcomings, FLIM FRET has recently been applied in a high-throughput, high-content approach for drug discovery (31). In this work, both high-throughput well-formats and multiplexed high-content setups were explored. A plate reader was modified with a Nipkow disc for optical sectioning and used to study the dimerization of HIV-1 gag proteins in a high-throughput manner. This approach of modifying technology designed for HTS to quickly measure detailed subcellular events is an example of the development of novel HCI technology. The application of FRET to study protein interactions or develop biosensors in high throughput could provide complementary insights into intracellular biological signalling processes during active cellular remodelling on biomaterials. For example, a FRET calcium biosensor (32) could be used to measure the calcium content of progenitor cells on a biomaterial library as a metabolic or differentiation readout of developing cells (e.g. neurons or cardiomyocytes) in stimulatory micro-niches.

Raman spectrometry/spectroscopy, a widely used analytical tool in physics, chemistry and materials science, has recently been explored for biomedical applications such as cancer diagnosis (33) and stem cell and regenerative medicine (34). Raman scattering, originating from the inelastic scattering of a photon, is intrinsically weak and barely detectable by the photomultiplier tubes or CCD cameras that are generally used in optical imaging and microscopy. As a result, several techniques have been developed to enhance Raman signals to an extent that is suitable for imaging applications, including coherent anti-Stokes Raman spectroscopy (CARS) imaging, which employs multiple photons to address the molecular vibration and produces a signal where the emitted waves are coherent with one another (35), surface-enhanced Raman spectroscopy (SERS) imaging which uses nanoparticles (sometimes called SERS tags or SERS dots) as signal enhancement agents (36), and single-walled carbon nanotubes (SWNT) imaging which utilizes its intense intrinsic Raman scatter at 1593 cm^{-1}(37). In the case of high-content/HTS where multiple biomarkers are usually needed to probe cell systems, SERS-based imaging is especially relevant because of its enhanced multiplexing capabilities. In general, nanoparticle-based SERS tags as labels for antibodies or other ligands offer several potential advantages over traditional fluorophores (38). First, like nanoparticle quantum dots, SERS tags are more photostable than conventional fluorescence dyes. Second, the nanoparticle resonance, and thus excitation wavelength, can be tuned by changing the size and shape of the nanoparticle, allowing the fabrication of near-infrared-excitable SERS tags that can reserve the space for fluorescent imaging at visible range. Finally, by using various Raman compounds that exhibit distinct Raman spectra, SERS tags can be prepared that have unique spectral signatures, thereby allowing a high level of multiplexing within a relatively narrow spectral range. As of now, SERS-based Raman imaging, although mostly demonstrated only in proof-of-concept experiments, could provide additional chemical/compositional information to aid future high-content screening research. In addition, it offers the possibility of acquiring many cellular parameters at the same time, for example, by attaching various molecular recognition agents to Raman tags to allow simultaneous probing of multiple biological events (cell survival, proliferation, apoptosis, differentiation, etc.).

6.7   Snapshot summary

- HCI has quickly become an important technique in the biomaterials research field.
- It is important to select the right imaging equipment for the material substrate and desired level of readout complexity.
- When designing the biomaterial platform for HCI it is important to make it suitable for the desired imaging technique.
- A robust assay is paramount to the success of a high-throughput experiment.
- Reporter cell lines are very useful for HCI.
- In general, location-based readouts are more robust than intensity-based readouts.
- HCI requires a large amount of computing power and expertise in bioinformatics.
- A good working image analysis pipeline requires a significant time investment.
- Both commercial and open-source software are viable, and the choice depends on the level of expertise of the user, the desired amount of flexibility and cost.
- Imaging techniques such as FRET FLIM and Raman, which are currently being adapted for high-throughput experiments, are promising techniques for biomaterials research.

Further reading

Bickle M. The beautiful cell: high-content screening in drug discovery. Analyt Bioanal Chem. 2010;**398**(1):219–26.

Treiser MD, Liu E, Dubin RA *et al*. Profiling cell–biomaterial interactions via cell-based fluororeporter imaging. Biotechniques. 2007;**43**(3):361–6, 8.

Kriston-Vizi J, Lim CA, Condron P *et al.* An automated high-content screening image analysis pipeline for the identification of selective autophagic inducers in human cancer cell lines. J Biomol Screen. 2010;**15**(7):869–81.

Carpenter AE, Jones TR, Lamprecht MR *et al.* CellProfiler: image analysis software for identifying and quantifying cell phenotypes. Genome Biol. 2006;**7**(10):R100.

Pernagallo S, Unciti-Broceta A, Diaz-Mochon JJ, Bradley M. Deciphering cellular morphology and biocompatibility using polymer microarrays. Biomed Mater. 2008;**3**(3):034112.

Gobaa S, Hoehnel S, Roccio M *et al.* Artificial niche microarrays for probing single stem cell fate in high-throughput. Nat Methods. 2011;**8**(11):949–55.

Mei Y, Saha K, Bogatyrev SR *et al.* Combinatorial development of biomaterials for clonal growth of human pluripotent stem cells. Nat Mater. 2010;**9**(9):768–78.

Inglese J, Johnson RL, Simeonov A *et al.* High-throughput screening assays for the identification of chemical probes. Nat Chem Biol. 2007;**3**(8):466–79.

Pal NR, Pal SK. A review on image segmentation techniques. Pattern Recogn. 1993;**26**(9):1277–94.

Kobel S, Lutolf M. High-throughput methods to define complex stem cell niches. Biotechniques. 2010;**48**(4):ix–xxii.

References

1. Masters BR. History of the optical microscope in cell biology and medicine. *Encyclopedia of Life Sciences*. Wiley; 2008.
2. Masters BR. The development of fluorescence microscopy. *Encyclopedia of Life Sciences*. Wiley; 2010.
3. Kobel S, Lutolf M. High-throughput methods to define complex stem cell niches. Biotechniques. 2010;**48**(4):ix–xxii. Epub 2010/06/24.
4. Bickle M. The beautiful cell: high-content screening in drug discovery. Analyt Bioanalyt Chem. 2010;**398**(1):219–26. Epub 2010/06/26.
5. Conrad C, Erfle H, Warnat P *et al.* Automatic identification of subcellular phenotypes on human cell arrays. Genome Res. 2004;**14**(6):1130–6. Epub 2004/06/03.
6. Bhadriraju K, Elliott JT, Nguyen M, Plant AL. Quantifying myosin light chain phosphorylation in single adherent cells with automated fluorescence microscopy. BMC Cell Biol. 2007; **8**:43. Epub 2007/10/19.
7. Pal NR, Pal SK. A review on image segmentation techniques. Pattern Recogn. 1993; **26**(9):1277–94.
8. Treiser MD, Liu E, Dubin RA *et al.* Profiling cell–biomaterial interactions via cell-based fluororeporter imaging. Biotechniques. 2007;**43**(3):361–6, 8. Epub 2007/10/03.
9. Yang J, Mei Y, Hook AL, Taylor M *et al.* Polymer surface functionalities that control human embryoid body cell adhesion revealed by high-throughput surface characterization of combinatorial material microarrays. Biomaterials. 2010;**31**(34):8827–38. Epub 2010/09/14.
10. Mei Y, Hollister-Lock J, Bogatyrev SR *et al.* A high-throughput micro-array system of polymer surfaces for the manipulation of primary pancreatic islet cells. Biomaterials. 2010;**31**(34):8989–95. Epub 2010/09/11.
11. Pernagallo S, Diaz-Mochon JJ, Bradley M. A cooperative polymer-DNA microarray approach to biomaterial investigation. Lab Chip. 2009;**9**(3):397–403. Epub 2009/01/22.
12. Fuchs F, Pau G, Kranz D *et al.* Clustering phenotype populations by genome-wide RNAi and multiparametric imaging. Mol Syst Biol. 2010;**6**:370. Epub 2010/06/10.

13. Bakal C, Aach J, Church G, Perrimon N. Quantitative morphological signatures define local signaling networks regulating cell morphology. Science. 2007;**316**(5832):1753–6. Epub 2007/ 06/26.

14. Treiser MD, Yang EH, Gordonov S *et al*. Cytoskeleton-based forecasting of stem cell lineage fates. Proc Natl Acad Sci USA. 2010;**107**(2):610–5. Epub 2010/01/19.

15. Kriston-Vizi J, Lim CA, Condron P *et al*. An automated high-content screening image analysis pipeline for the identification of selective autophagic inducers in human cancer cell lines. J Biomol Screen. 2010;**15**(7):869–81. Epub 2010/06/16.

16. Inglese J, Johnson RL, Simeonov A *et al*. High-throughput screening assays for the identification of chemical probes. Nat Chem Biol. 2007;**3**(8):466–79. Epub 2007/07/20.

17. Kozak K, Bakos G, Hoff A *et al*. Workflow-based software environment for large-scale biological experiments. J Biomol Screen. 2010;**15**(7):892–9. Epub 2010/07/14.

18. Conrad C, Wunsche A, Tan TH *et al*. Micropilot: automation of fluorescence microscopy-based imaging for systems biology. Nature Methods. 2011;**8**(3):246-U89.

19. Pau G, Fuchs F, Sklyar O, Boutros M, Huber W. EBImage – an R package for image processing with applications to cellular phenotypes. Bioinformatics. 2010;**26**(7):979–81. Epub 2010/03/27.

20. Abramoff MD, Magelhaes PJ, Ram SJ. Image processing with ImageJ. Biophotonics International. 2004;**11**(7):36–42.

21. Carpenter AE, Jones TR, Lamprecht MR *et al*. CellProfiler: image analysis software for identifying and quantifying cell phenotypes. Genome Biol. 2006;**7**(10):R100. Epub 2006/ 11/02.

22. Mayr LM, Bojanic D. Novel trends in high-throughput screening. Curr Opin Pharmacol. 2009;**9**(5):580–8. Epub 2009/09/25.

23. Joseph J, Seervi M, Sobhan PK, Retnabai ST. High-throughput ratio imaging to profile caspase activity: potential application in multiparameter high content apoptosis analysis and drug screening. PLoS One. 2011;**6**(5):e20114. Epub 2011/06/04.

24. Mei Y, Saha K, Bogatyrev SR *et al*. Combinatorial development of biomaterials for clonal growth of human pluripotent stem cells. Nat Mater. 2010;**9**(9):768–78.

25. Pozarowski P, Holden E, Darzynkiewicz Z. Laser scanning cytometry: principles and applications. Methods Mol Biol. 2006;**319**:165–92. Epub 2006/05/25.

26. Pernagallo S, Unciti-Broceta A, Diaz-Mochon JJ, Bradley M. Deciphering cellular morphology and biocompatibility using polymer microarrays. Biomed Mater. 2008;**3**(3):034112. Epub 2008/08/19.

27. Khan F, Tare RS, Kanczler JM, Oreffo RO, Bradley M. Strategies for cell manipulation and skeletal tissue engineering using high-throughput polymer blend formulation and microarray techniques. Biomaterials. 2010;**31**(8):2216–28. Epub 2010/01/09.

28. Unadkat HV, Hulsman M, Cornelissen K *et al*. An algorithm-based topographical biomaterials library to instruct cell fate. Proc Natl Acad Sci USA. 2011;**108**(40):16565–70. Epub 2011/09/29.

29. Dalby MJ, Gadegaard N, Tare R *et al*. The control of human mesenchymal cell differentiation using nanoscale symmetry and disorder. Nat Mater. 2007;**6**(12):997–1003. Epub 2007/09/25.

30. Lovmand J, Justesen J, Foss M *et al*. The use of combinatorial topographical libraries for the screening of enhanced osteogenic expression and mineralization. Biomaterials. 2009;**30**(11):2015–22. Epub 2009/01/31.

31. Kumar S, Alibhai D, Margineanu A *et al*. FLIM FRET technology for drug discovery: automated multiwell-plate high-content analysis, multiplexed readouts and application in situ. Chemphyschem. 2011;**12**(3):609–26. Epub 2011/02/22.

32. Mank M, Reiff DF, Heim N *et al.* FRET-based calcium biosensor with fast signal kinetics and high fluorescence change. Biophys J. 2006;**90**(5):1790–6.

33. Zhang Y, Hong H, Cai W. Imaging with Raman spectroscopy. Curr Pharm Biotechnol. 2010;**11** (6):654–61. Epub 2010/05/26.

34. Schulze HG, Konorov SO, Caron NJ *et al.* Assessing differentiation status of human embryonic stem cells noninvasively using Raman microspectroscopy. Analyt Chem. 2010;**82**(12):5020–7. Epub 2010/05/21.

35. Evans CL, Xie XS. Coherent anti-Stokes Raman scattering microscopy: chemical imaging for biology and medicine. Annu Rev Analyt Chem (Palo Alto Calif). 2008;**1**:883–909. Epub 2008/07/19.

36. Schlucker S. SERS microscopy: nanoparticle probes and biomedical applications. ChemPhysChem. 2009;**10**(9–10):1344–54. Epub 2009/07/01.

37. Hong H, Gao T, Cai W. Molecular imaging with single-walled carbon nanotubes. Nano Today. 2009;**4**(3):252–61. Epub 2009/06/01.

38. Nolan JP, Sebba DS. Surface-enhanced Raman scattering (SERS) cytometry. Methods Cell Biol. 2011;**102**:515–32. Epub 2011/06/28.

39. Liu E, Treiser MD, Patel H *et al.* High-content profiling of cell responsiveness to graded substrates based on combinatorially variant polymers. Comb Chem High Throughput Screen. 2009;**12**(7):646–55. Epub 2009/06/18.

40. Kiel MJ, Morrison SJ. Uncertainty in the niches that maintain haematopoietic stem cells. Nat Rev Immunol. 2008;**8**(4):290–301.

41. Gobaa S, Hoehnel S, Roccio M *et al.* Artificial niche microarrays for probing single stem cell fate in high throughput. Nat Methods. 2011;**8**(11):949–55. Epub 2011/10/11.

7 Computational analysis of high-throughput material screens

Marc Hulsman, Liesbet Geris and Marcel J. T. Reinders

Scope

Computational and statistical tools play an important role in materiomics, to provide insights in the underlying processes that allow certain materials to outperform other materials. In this chapter, we discuss numerous methods that allow the analysis of materiomics data. Specifically, we describe the use of statistical tests, ranking and data mining approaches, model learning and testing, as well as experimental design and the exchange of experimental results. Also, we review some of the important publications in this field from the past 15 years, organizing them according to the type of material descriptors that were used.

7.1 Basic principles of data analysis

Computational methods play an ever more important role in the study of material function. Partly, this is due to the increased scale of the experiments being performed, with an accompanying need for automated analyses. But the move from low-throughput towards high-throughput experiments entails more than just testing more materials simultaneously. The extra information these experiments produce is slowly catalysing a transition to a more rational approach to material discovery, in which not just material screening plays a role but also material modelling. Materials and their environments are approached as systems that can be modelled and thus explored *in silico*. This 'systems approach to material research' has been termed materiomics. This transition is certainly needed given the size of the materiome that one wants to explore: many material parameters can be varied and combined into a practically infinite palette of combinations. This far surpasses even the reach of high-throughput screenings. The question that will be addressed in this chapter is: how can we efficiently make use of our capability to perform high-throughput experiments, to explore and characterize such a large search space?

The materiomics approach entails that we are looking for model $y_m = f(x_m)$, which expresses material performance y_m (e.g. material m's capacity to promote protein adsorption) in terms of physico-chemical descriptors $x_m = (x_{1,m}, \ldots, x_{n,m})$ of the material

Materiomics: High-Throughput Screening of Biomaterial Properties, ed. Jan de Boer and Clemens van Blitterswijk. Published by Cambridge University Press. © Cambridge University Press 2013.

and its environment. Such a model subsequently allows one to improve performance by making rational changes to a material or its environment.

In earlier chapters, the various physico-chemical descriptors and material perform-ances, as well as the computational analysis to automatically extract them (e.g. image analysis of high-throughput data), have been discussed. Here, we first describe methods that can be used to analyse measurements of material performance y_m. While these methods are also applied in low-throughput settings, the high-throughput setting leads to new challenges. Next, we discuss modelling methods, which allow one to construct a model $f(x_m)$ predictive of material performance. Finally, we look at experimental design, that is, choosing which materials to test and descriptors to use (x_m).

7.1.1   Statistical inference: y

Box 7.1 Statistical testing

- Hypothesis: assumption about the state of a system and/or the factors that influence the measurement process.
- Hypothesis test: determine the likelihood of the current data (p-value), given that a certain hypothesis is correct.
- Hypothesis rejection: for low p-values (e.g. below 0.05), one can reject the hypothesis.
- ANOVA test: assumes the hypothesis that all materials perform similarly.
- t-test: assumes the hypothesis that two materials perform similarly.

Testing for material effects

Does varying the material design affect the material performance? This is one of the first questions that is asked when analysing a materiomics dataset. For example, can we show that by varying surface structure, one can influence cell proliferation? Such questions can be answered through the well-known ANOVA test (1). Given the assumption that the actual material performances for k different materials $m_1, \ldots, m_k$ are equal, i.e. $y_{m1} = y_{m2} = \cdots = y_{mk}$, the ANOVA test reports a p-value, representing how likely it is that we measure material performance differences that exceed those that have been observed. That is, one tests whether normal measurement variance can explain the observed performance differences. The equality assumption is rejected if the p-value falls below a threshold (usually set at 0.05 or 0.01), allowing one to conclude that there is indeed a significant material effect on the performance y by (one of) the materials $m_1, \ldots, m_k$.

It is important to consider that the opposite is, however, not true: a failure to reject the equality assumption does not imply that there is no material effect. The effect might instead be hidden by large amounts of natural and/or technical variation. Also, note that an ANOVA test can only be performed if one has replicate measurements of the performance (y) for (at least some of) the materials[1]. The number of required replicates depends on the effect size (i.e. the expected differences in material performance) as well as the measurement error and the number of materials. As the number of materials is fairly large in a high-throughput setting, the number of required replicates for this test is normally relatively low.

Material comparison

Determining whether any of the tested materials works better than some reference, state-of-the-art material is often one of the main goals of an experiment. To answer this, one can employ a standard one-tailed Student's t-test. Such a test determines for a material m_i whether the 'perform equally or worse' assumption $y_{m_i} \leq y_{m_{\text{reference}}}$ can be rejected. In a high-throughput setting where this test is performed separately for each material m_i ($i = 1$ to N), this leads to what is known as the multiple testing effect (2). For a single test, one can expect to find for random data that there is one incorrectly assigned significant positive result for every 20 tests (given a p-value threshold of $\alpha = 0.05$).[2] This is usually deemed acceptable. However, when applying this test to $N = 1000$ materials, this can translate into finding (just by chance) 50 materials that 'significantly' outperform the reference material (so-called false positives). An often used, although highly conservative, method to correct for this effect is Bonferroni correction (2). A new p-value threshold[3] $\alpha_{\text{bonferroni}} = \frac{\alpha}{N}$ is calculated, where N is the number of tests that are performed. With this new threshold, the chance of false significant results is now less than α for all tests combined (known as the family-wise error rate, FWER).

However, this is usually more stringent than needed, especially when the study is exploratory, i.e. looking for promising leads. In such cases, the false discovery rate (FDR) is more suitable. This is the expected ratio of false positives one accepts in a set of top leads. For example, say one has tested 1000 materials. An FDR cut-off of 0.1 might result in 20 candidates, of which one can then expect only 2 (10% of 20) to be false positives. These leads can then be further validated in a follow-up experiment. A well-known FDR method is the Benjamini–Hochberg procedure (3).

Besides these approaches, various other comparison methods have been described in the literature, where this problem is known as multiple comparisons with a control (MCC) (4). Note that multiple testing corrections do have important consequences for study design. Possibly counterintuitively, with more tested materials one will also require more replicates per material. This is due to the more stringent control for false positive results. More statistical power (more replicates) is thus required to consider a test significant.

---

[1] An alternative is to test for a relation between material descriptors x and y, e.g. using the correlation-test or the F-test. This accomplishes the same goal and does not explicitly require replicate measurements. These tests do, however, require assumptions on the relations between the different materials with respect to the performance (e.g. linear relations between material descriptors and performance), which usually is exactly the information that is missing. Also, while this test does not need replicates, it does require a large number of tested materials.

[2] A p-value of 0.05 means that there is a probability of 1/20 that a material i with equal performance to the reference material shows a performance of y_{m_i} or larger.

[3] Strictly speaking, Bonferonni correction corrects the p-value and not the cut-off level.

So, where in a low-throughput setting three replicates might have been enough to prove that a certain material performs better than the reference, in a high-throughput setting one might need six replicates to prove the same.

Material ranking and selection

Another goal of materiomics studies is the discovery of hit materials, i.e. those materials that are the most promising in terms of performance. To this end, materials can be ordered on their observed performance (averaged across replicates). Such an approach is, however, prone to outlier effects: replicates with an extreme outlier score will strongly affect the material ranking. Highly ranked materials could thus be in such a position only by virtue of only one highly performing replicate.

There are various methods available that, based on replicate information, control for such effects. One is the MCC approach described earlier: materials can be ordered according to their statistical score (p-value), obtained from a performance comparison with a reference material. Such scores are more robust to outlier effects, as the variance of the replicate measurements, as well as the number of replicates, is taken into account. Essentially, one orders not on observed performance, but rather on confidence in a high performance.

Irrespective of the ranking method used, however, the best scoring material does not actually have to be the true best performing material. Especially if there are few replicates and many materials, measurement errors may lead to incorrect rankings. As a solution to this problem, MCC methods have been adapted, to perform instead a 'multiple comparison with the best' (MCB) (4). This type of method allows one to select all tested materials that could potentially be the true best performer. Remaining materials, with a performance confidence region below the best observed performance, can be excluded from further testing, as the experiment has persuasively shown that they are not the best performers.

To select the true best performer, follow-up experiments can then be performed on the hits. Two questions remain: what is the minimum performance difference between materials that we still care about (necessary for the specification of the confidence region)? And, given that, how many extra replicates do we need in follow-up experiments? If one specifies the answer to the first question, ranking and selection (RS) methods can be used to answer the second question (5).

Besides the hit materials, useful knowledge can still be obtained by exploring performance differences in the remaining tested materials. To statistically determine whether any two materials can be distinguished from each other in terms of performance, an ANOVA post-hoc test can be employed. A well-known one is Tukey's range test, which performs all pairwise comparisons between materials, while controlling for multiple testing (e.g.(6)).

Multiple material performances

Material performance cannot always be expressed as a single value. For example, when determining the effect of material composition on cell morphology, both cell circumference and area are relevant. It is possible to test for multivariate material effects (Section 7.1.1) using a multivariate version of ANOVA (MANOVA). This is not the case for material rankings, as it is not *a priori* defined how multiple performances $y_m = \left(y_{1,m}, \ldots, y_{n,m} \right)$ (where $y_{n,m}$ is the nth performance measure for material m) should be combined into one ranking (what is the best material?). This problem can be handled either by using techniques that consider multiple optimum solutions (known as Pareto optimality), or by converting the multivariate problem into a univariate one (e.g. using area/circumference) (7).

Exploring material responses

Up to now the discussion has centred on asking well-defined questions. It can also be fruitful to explore the data for other interesting patterns, an activity known as data mining or exploratory data analysis. Such exploration is usually accomplished by visualizing the measurement results. While such a visualization is relatively simple for a single measured property (e.g. a histogram) or two measured properties (e.g. a scatter plot, with the two properties along the two axes), it becomes less straightforward in cases where three or more measured properties have to be plotted simultaneously. For example, one might want to explore 'cell morphology', which could reveal that cells differentiate into two or more distinct morphologies. Such morphologies are normally characterized by a large number (10+) of properties. Plotting along so many properties (dimensions) is not trivial. One way to visualize such data is to decrease the number of dimensions, in such a way that one still captures the distribution of the data in the original space. This can be accomplished through the use of dimension scaling methods. These methods transform a set of material properties into a smaller set of derived properties. A well known method is principal component analysis (PCA), but there are others, such as independent component analysis (ICA) and multidimensional scaling (MDS) (8). Note that dimensional scaling is a lossy process: not all information contained in the original set of properties is (normally) kept. The methods differ in the way in which they decide what information is important, and how this information should be distributed over the set of derived properties. PCA seeks derived properties that are decorrelated, ordering them on the amount of data variance they explain. The first components are thus usually the most important (see Figure 7.1 for an example).

ICA, in contrast, seeks derived properties that are independent and does not order properties. ICA is often used to perform blind source separation (with the classical example being speech source separation at a cocktail party). Kardamakis, Mouchtaris and Pasadakis (9) used this technique to determine refinery fractions of gasoline mixtures, using only infrared spectra. Finally, MDS seeks to optimize the distances between the materials, in the new lower-dimensional space, such that these distances correspond as closely as possible to the distances in the original (high-dimensional) space. For all these methods, interpretation of the results in terms of the original properties (dimensions) is difficult.

Another way to explore data is to group the data using clustering techniques such that materials with similar performances group together. Two commonly used algorithms are k-means clustering and hierarchical clustering (10). The first algorithm seeks to optimize, for a predefined number of clusters, the distance of each material from its closest cluster centre, by moving around candidate cluster centres. Hierarchical clustering on the other hand iteratively groups the most similar materials together into

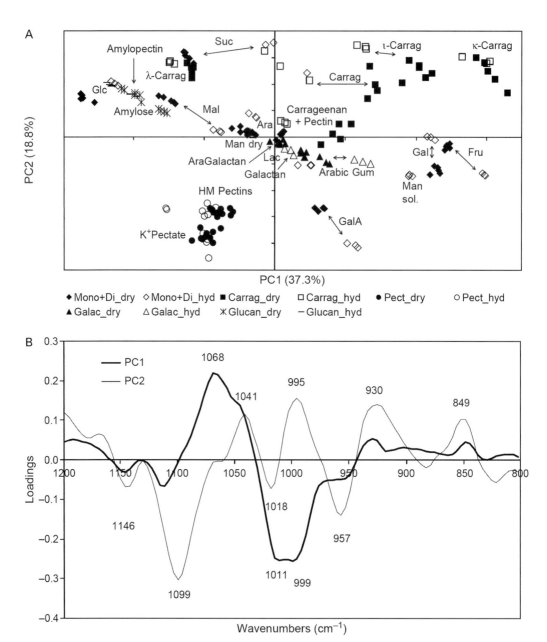

Figure 7.1 Distinguishing sugars in dry and hydrated environments, using FTIR spectra. A, A scatter plot of the spectra of standard mono-, di- and polysacharides, represented by the first two PCA dimensions. B, The loadings of both PCA components (the loading indicates the weight with which each original property (infrared wavenumber, which is the reciprocal of wavelength) has contributed to a PCA component). Reproduced with permission from (102).

larger clusters. The result is a tree of clusters. This latter form of clustering is also often applied to create the well-known transcriptomics heat-map visualizations (e.g. (11)), where one performs a clustering not only on genes, but also on samples. Note that there is no straightforward statistical test that can determine the significance of a clustering, (101) at least not without an external reference frame such as a predefined grouping of materials to which the clustering can be compared. Clustering scoring methods (e.g. the Davies–Bouldin index, silhouette methods, cluster stability)(12) can, however, be used to compare different clusterings, for example to choose the right number of clusters. When interpreting a clustering result, one should keep in mind that clustering algorithms always return a clustering, even if there is no real support for clusters in the data. The results of a clustering should therefore be inspecting by visualizing them, for example using a heat map or a (MDS) scatter plot.

Box 7.2 Classic experiment

Predicting fibrinogen adsorption using theoretically derived compound descriptors and surrogate models

It is well known that the relation between molecular structure and molecular function can be complex. Linking a molecular structure directly to a function such as fibrinogen adsorption would mean that the model learning algorithm has to infer such a relation from scratch, based on relatively limited data.

In Gubskaya *et al.* (13), theoretical modelling techniques were employed to partially cover this gap, by using simulations to derive molecular properties from the molecular structure. Specifically, simulations were performed to calculate 3D polymer descriptors, which were subsequently used as input for the machine learning algorithm (Figure 7.2). The paper is a classic example of how various techniques can be used to optimize the predictive ability of surrogate models.

Briefly, for 45 polymers, 3D molecular structures were constructed by performing molecular dynamics (MD) simulations using MacroModel software (14). In these simulations, an empirical atom force field was used to determine the optimal conformation of the atoms. The quality of these predictions was assessed by repeating the minimization for different initial structures.

Given this 3D model of the molecular compounds, the DRAGON software was used to calculate 3D molecular descriptors. In total, 721 descriptors were calculated for each polymer, describing their geometry, topology and other chemical information. However, with many more descriptors (721) than materials (45), machine learning algorithms would surely over-fit. Therefore, in the next step, a similar approach as developed in (15) was used to reduce the set of descriptors to only the most relevant ones. The study showed that various classes of descriptors are relevant, including distances between oxygen atoms and between oxygen and nitrogen atoms, which the authors related to the importance of electrostatic interactions and hydrogen bonding.

Given these descriptors, in the next step, an artificial neural network (ANN) was trained to predict fibrinogen adsorption. It was shown that descriptors based on 3D structures obtained using an MD simulation outperform techniques in which the structure is obtained through energy minimization. Also, the performance of the 3D-descriptor model compared favourably with models based on 2D descriptors.

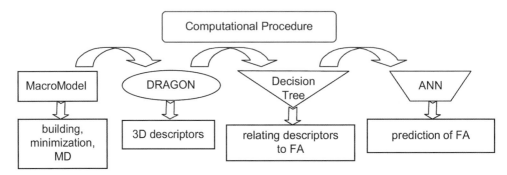

Figure 7.2   Overview of method used by Gubskaya *et al.* (13) to predict fibrinogen adsorption (FA). A theoretical model is used to define 3D descriptors. Subsequently, the most relevant descriptors for fibrinogen adsorption are determined, and used to build a predictive model. Reproduced with permission from (13).

7.1.2   Model learning: $f(x)$

With modelling, the relation between material performance y and material descriptors x is investigated. In an ideal situation, such a material performance model can be constructed from first principles, i.e. by quantum mechanical simulations. Such simulations, however, quickly become intractable when there are more than 10^2 atoms involved (16). To address this limitation, multiscale simulations have been developed, in which lower-scale simulations are used to parameterize higher-scale simulations. For example, a low-scale quantum mechanical simulation might enable a force-field simulation, which in turn might enable a molecular dynamics simulation, which in turn might enable a simulation representing the material as a continuum (Figure 7.3) (17).

With the increase in scale, not only larger stretches of material can be simulated (from angstroms to metres), but also longer time periods (from picoseconds to minutes). A nice example of this approach is given in a recent article analysing the behaviour of spider webs under stress (18).

Such simulations, however, have their limitations: many problems are still intractable (e.g. protein folding) and knowledge of the material composition, structure and environment is required. The latter issue becomes an especial concern when 'black boxes' such as biological cells are involved (e.g. when investigating cell attachment). This is where

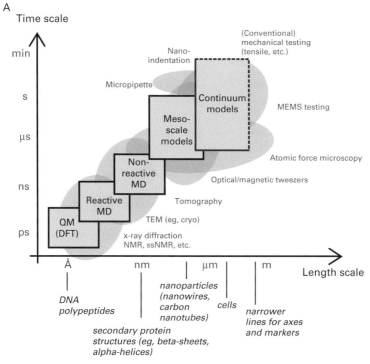

Figure 7.3   A, Length and time scales covered by different material modelling and measurement methods. The lack of modelling methods that can cover the whole time and length range indicates that approaches are needed that combine multiple modelling scales and methods. B, Large-scale biological structures consist of repetitive lower-scale structures (e.g. alpha helices, beta sheets). This hierarchical structure allows for the efficient combination of multiple modelling scales, as one can make use of this repetitive structure to reduce computational effort for the lower scales. MD, molecular dynamics; QM, quantum mechanical modelling, DFT, density functional theory; MEMS, microelectromechanical systems, ssNMR, solid-state nuclear magnetic resonance. Reproduced with permission from (17).

empirical models come in: instead of beginning from known principles (bottom-up), one might start from the experimental results (top-down). When enough information is available on the problem domain to construct at least an initial simulation model, one can use these experimental results to fill in (fit) the model parameters. Otherwise, machine learning tools can be used to find, *ab initio*, a plausible underlying model that can explain the experimental results. The latter types of models are the most frequently used, and known in biomaterial literature as surrogate models (19). We discuss both forms of modelling in the subsequent sections.

Model choice
Learning surrogate models (20) comes down to finding the function f in $y = f(x)$, using what are called training examples. Such training examples, each representing a material m_i, contain both the outcome measurement y_{m_i} (called a label) and the n input material

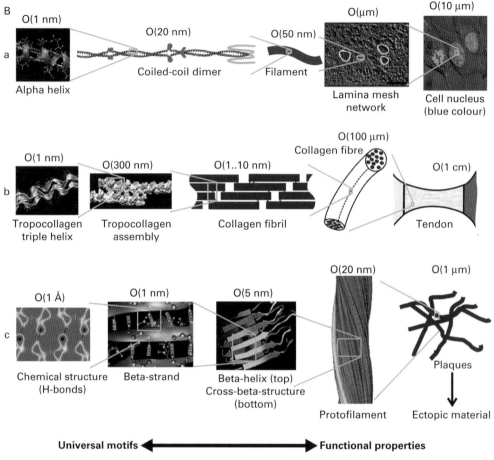

Universal motifs ◀━━━━━━━━━━━━▶ Functional properties

Figure 7.3   (cont.)

descriptors $x_{m_i} = \left(x_{1,m_i}, \ldots, x_{n,m_i}\right)$ (called features). Learning algorithms use these train-ing examples to search for a model which accurately predicts the label, given the features. This process is known as training.

Considering the material performance, there are two different cases: one in which the label y has a continuous value (e.g. the fraction of the area filled with bone), and one in which it has a discrete value (e.g. 'bone-forming' or 'non-bone-forming'). Machine learning problems corresponding to these two cases are respectively called regression and classification. The name 'classification' here reflects the fact that a set of objects with a common label is called a class, e.g. the class of materials that induce bone-forming.

For both problem types, a plethora of learning methods is available. An important characteristic to consider is the amount of complexity these different algorithms allow in a model. The simpler algorithms are linear, i.e. they look for models with linear responses between inputs and outputs. For example, linear regression fits a model-function $y = \beta_0 + \beta_1 x_1 + \cdots + \beta_n x_n$ to the data (the algorithm has to estimate the β_k values from the training examples). More complex are algorithms that allow for non-linear responses. Frequently used non-linear algorithms are artificial neural networks (ANN), decision trees, random forests and support vector machines (SVM) (21), all available in versions for classification and regression.

Choosing the algorithm that is capable of finding the most intricate, complex, models is not necessarily the best strategy. With more freedom, an algorithm will be 'tempted' to learn to reproduce just the individual training data responses, instead of generalizing to the true underlying process. This effect is called 'over-fitting'. One could compare this to a student who has learned the answer to a list of exam questions (overtraining on the training examples), but is incapable of answering variations on these questions (generalization to other examples). One solution to this problem is to significantly increase the number of training examples, so that the model will be incapable of capturing the answers to each of the individual examples, forcing it to generalize across the examples. This approach is not always viable, as generating these extra examples can be costly. Another solution (and most used) is to reduce the allowed model complexity in such a way that the model is not capable of reproducing every individual training example, again forcing the model to generalize. Clearly, the model should still be complex enough to capture the underlying relations in order to be able to have enough generalization power. Note that most learning algorithms have parameters through which one can set the allowed model complexity. Constraining or penalizing these parameters is then the true art of creating proper models.

Model assessment

An essential step in model learning is model testing. As we have seen, models learned from experimental data are not necessarily correct. One can easily end up with a model that only works on training examples. Testing a model requires unseen test examples: materials for which x and y are known, but that are not used during the learning phase. The quality of a surrogate model is indicated by its predictive capabilities on these test examples, by comparing the predicted response of the model, $f(x)$, with the measured value y. It is crucial that the model is not tested with examples used during learning: the reported performance would then be positively biased, as the model has been optimized to predict these example materials.

Given the expenses associated with material testing, withholding test data just for performance estimation might seem wasteful, as one generally obtains better models by using more training data. Still, a large enough set of test data is needed in order to obtain stable performance estimates, as without such an estimate the model is essentially worthless.

To have one's cake (testing data) and eat it too (training data), one can make use of a scheme called cross-validation (22). Briefly, one splits the data into K partitions. Then, each partition is in turn used as the test set, while the remaining $K-1$ partitions are used to train the model. This results in K performance numbers, one for each combination of test/training set. These performances are subsequently averaged to

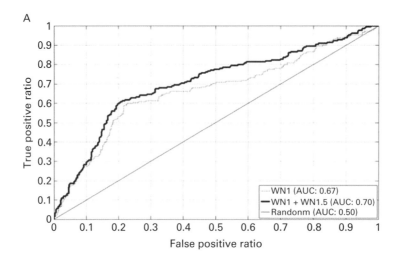

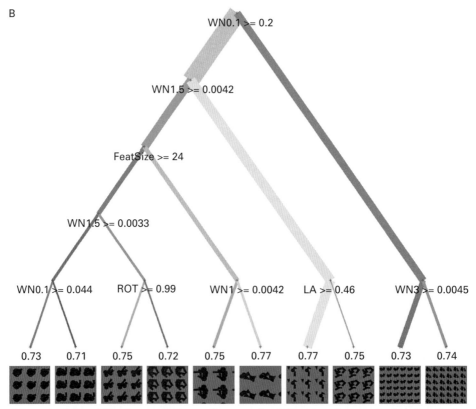

Figure 7.4   **A**, Cell proliferation prediction performance for different surface topologies, described using a ROC curve. Shown here is the effect of choosing two different sets of features (WN1 and WN1 + WN1.5) on the performance. The WN features indicate the repetition of

arrive at a final performance estimate (the variance gives an estimate on the variability of the performance for this size of the training set). When K is equal to the number of available examples, this scheme is known as leave-one-out cross-validation[4].

To characterize the difference between predicted and measured material performance, various measures can be used. For classification, one of the most straightforward measures is the accuracy: the ratio of correct class predictions for test materials (estimated by comparing the class label predicted by the learned model with the actual known class label). Most classifiers, however, do not just assign a class, but also describe how certain they are of that classification.

This is useful information, as it allows one to distinguish high-quality predictions from low-quality ones. To visualize how this information relates to the actual measurements, a commonly used tool is the receiver operator characteristic (ROC) curve (Figure 7.4A) (23).

Given a specific class of interest (e.g. 'bone-forming'), the ROC is generated by selecting a certain threshold for the certainty with which the classifier outputs a positive class (in this case 'bone-forming'). By changing this threshold one can generate different operating points, i.e. changing the amount of examples that are considered positive. By generating all possible operating points (sweeping from calling only the most certain example positive to calling all examples positive), a curve is created. To draw the ROC, each operating point is plotted with respect to the number of false positives (x-axis) and true positives (y-axis). The resulting ROC curve is usually summarized into a single score, the area under the curve (AUC) score, after normalizing both axes of the ROC graph to the [0,1] domain. Such an AUC score is 0.5 for a classifier which randomly assigns classes, and 1.0 for a perfect classifier. A p-value can be attached to the AUC score through its connection with the Mann–Whitney U test (24).

The previous test scores can be used when classifying examples. When the model generates a continuous output instead of a class label, the Pearson correlation can be used to evaluate performance. Specifically, it is used to determine the correlation between the known performances (y values) and the performances predicted ($f(x)$) by the learned regression model. Similar to the AUC score, a p-value can be attached to this score. This makes the measure also suitable for testing possible relations between two material properties.

Caption for Figure 7.4 (cont.)
the surface patterns for certain frequencies (indicated by x). The 'random' line indicates the performance that can be expected from a completely random classification. Reprinted with permission from (84). B, Decision tree, predicting nucleus shape, based on surface features (which are represented by terms such as ROT, TA, etc., each describing a specific aspect of the surface topography, the details of which are not important in the current context). The thickness of the lines indicates the number of surfaces that follow a certain path. Below the tree, we show the predicted values, as well as representative surfaces for each decision path.

---

[4] Generally, choosing the proper K is not trivial, owing to the bias-variance problem in model learning/testing.

Role of material descriptors

Knowing the role that each material descriptor x_i plays in a model is important for two reasons. First of all, it helps model interpretation, indicating which material descriptors are most predictive. It is also extremely helpful for controlling the model complexity during model learning (to avoid the over-fitting). Two approaches can be followed (25). One approach is to reduce the dimensionality of the learning space by dimensionality reduction methods. The approaches discussed earlier (PCA, ICA, MDS) are unsupervised (i.e. they do not use information about the label y). Using knowledge about how well certain descriptors predict outcome allows one to reduce dimensions in a well controlled manner. A well-known strategy for classification is Fisher discriminant analysis (22). For regression, partial least squares (PLS) regression can be used (22). Although these 'feature extraction' methods reduce dimensions, they do so by transforming the feature space by combining the original features, so that eventually the resulting models are hard to interpret in the original features.

A good alternative is 'feature selection', in which an informative subset of material descriptors is selected, using various search schemes (filtering, forward/backward, wrapper etc (26). and (supervised) material descriptor importance scores. A well-known learning algorithm with built-in feature selection is the decision tree algorithm (Figure 7.4B). It can be used both for regression and classification, and the resulting models are especially well-suited for interpretation. The learned models have the form of a tree, in which at each fork a decision is based on the value of certain relevant material descriptor. Leaves (i.e. end-points) of the tree contain the predictive values $f(x)$. Predictions are obtained by 'walking', for a certain material, up the tree from root to leaf. In this type of model, one can thus directly see how the different material descriptors each play a role in a certain prediction result. To select the model complexity (for example by selecting a subset of features, either using feature extraction or selection) again requires a model testing procedure. Again, it is crucial that examples used for model complexity selection are not also used for assessing the performance of the final model. Therefore an additional set of validation examples should be used. Consequently the initial data should be split into training samples (to train the model), validation samples (for model selection) and test samples (for model assessment). Both the test and the validation samples can be selected using a cross-validation scheme (resulting in a cross-validation scheme within a cross-validation scheme for the validation samples) (27).

Theoretical models

Theoretical models are, in contrast to surrogate models, not data-driven, but rather hypothesis-driven. That is, biological hypotheses are encoded into a model, after which model validation is used to assess the explanatory power of these hypotheses in predicting an actual observed system response. If the model performance is unsatisfactory, one can do this in a cyclic fashion, in which the model is adapted in response to the validation, and then tested again. Note that the same concerns that apply to surrogate model validation (i.e. not using data for testing that has also been used for learning) also apply to theoretical models. Both continuous and agent-based techniques are regularly used to construct theoretical models. The former type of modelling describes a continuous spatial domain, in which diffusion, advection and reaction processes are usually described through differential equations. The latter type divides the model space into different agents, each affecting their own neighbourhood. Often, agent-based models are constructed by dividing a continuous space into a grid, in which each grid element can, for example, describe a material unit or biological cell. A popular approach here has been the cellular Potts model (28), allowing one to model processes such as cell movement, growth and adhesion. Open-source software (CompuCell3D (29)) has been developed to support the implementation of such models.

Frequently, continuous and agent-based techniques are combined to construct models (e.g. (30)). These techniques can be complemented by more specialized approaches, e.g. modelling the rheological properties of blood, the blood flow through a network, and its effect on the wall shear stress (31).

As well as this horizontal model integration, many researchers are also using vertical model integration, in which models at different scales (multiscale) are combined, as has been discussed at the beginning of this section. This enables one to overcome significant computational bottlenecks.

7.1.3   Experimental design: *x*

The search for better materials is a typical optimization problem. Visualizing it as a landscape of material designs, in which the height of the landscape indicates material performance, the goal is to find the highest hill top. To be able to learn models, it is essential that each material is described in the same way: this common description acts as a 'coordinate system' in the design landscape, thereby allowing machine learning algorithms to exploit material similarity.

Material libraries

A collection (or continuum) of materials with common descriptors is usually referred to as a library. With n material descriptors that can influence a material design, each tested at l different levels, a complete library should contain l^n materials. This, however, is prohibitively large even for many high-throughput setups. Using techniques from the field of design-of-experiments (DOE), an optimal (small) subset of experiments can be selected for testing.

These experimental designs specify which material descriptor value combinations need to be tested, to obtain enough data to learn a model of a certain specified complexity. In practice, either linear (e.g. Placket–Burmann designs) or quadratic (e.g. Box–Behnken designs) models are used (32).

Compared with a uniform distribution of the samples over the material space, the use of these designs appears mainly advantageous for very low sample sizes (33, 34). Furthermore, descriptor values (e.g. hydrophobicity) do not always directly translate back into a material design, making it difficult to obtain the test materials that are prescribed by the experimental design. Therefore, algorithms (diversity methods) have been developed that take an available library as starting point. Subsets of materials are selected that optimally cover the material design landscape, by minimizing the similarity between the selected materials (35).

Unfortunately, one of the most pressing questions in designing an experiment, namely the number of materials that should be tested to obtain a good model, is hard to answer. The answer is strongly dependent on the strength of the relations between material descriptors, the (non-)linearity of the required model relations and the noise in the measurements. Generally, therefore, one makes as many measurements as experimentally feasible.

Sequential designs and optimization algorithms

Although the goal in the previous paragraph was to distribute the tested materials optimally across the material landscape, most of this landscape will be uninteresting in terms of performance. Sequential experimental designs can take advantage of this fact, by balancing landscape exploration with a more detailed investigation of interesting regions. Such an approach can lead to large experimental savings.

There are several algorithms that can support sequential experimental designs. A replicate-based approach was already mentioned in Section 7.1.1. Model-based approaches are also available, in the form of numerous optimization algorithms (36) and 'active learning' methods (37). These methods suggest in every iteration (a batch of) new materials which, if tested, are expected to be the most informative with respect to the goal.

Exchange of experimental results

Achieving the advantages of sequential experiments across studies or even labs requires the exchange of experimental results. Databases are tools that make this possible. Such databases can flourish only if extensive standardization is embraced. Agreements are necessary on which experimental conditions, material designs and material properties y and x are measured and/or represented, and how this is done. Within the materials field, researchers have explored designs of such a database for polymers (38). Popular technologies for these types of databases are based on the Semantic Web (39) framework, where a common information representation contract, called ontology (40), is used to enable data exchange and integration. Such ontologies are also widely used throughout bioinformatics (41).

7.2   Computational analysis in materiomics research

Faced with the problem of an almost infinitely large materiome, and inspired by immunobiology and drug screening approaches, material scientists have developed a large array of high-throughput material testing techniques. Starting with studies into material functions such as superconductivity (42) and luminescence (43), initial focus has been on the development of experimental techniques. Three years after the pioneering work in 1995 (42), the field had already grown enormously, with three new journals, and more than 200 relevant publications reviewed in (44). Many of these pioneering studies, however, took a more or less manual approach to the analysis of their results.

Nevertheless, with the availability of the new experimental techniques, adoption of new analysis techniques has followed quickly. A prime example of this is the polymer materials field, where the creation of large material libraries became possible early on through the use of combinatorial chemistry, which has led to the development of various machine learning methods for predicting polymer functions. However, other materials, such as ceramics, catalysts and (designed) structural materials, have also been investigated. Numerous methods have been applied, often centred around different forms of material descriptors, which we have taken as a way to organize this review.

7.2.1   Spectroscopy descriptors: descriptor reduction and interpretation

That the application of new analysis techniques was not contingent on the development of high-throughput experimental techniques was shown by Pérez-Luna, Horbett and Ratner (45). In this study, spectroscopic and contact angle measurements of 21 polymers were related to fibrinogen adsorption and retention. As spectroscopic measurements result in numerous descriptors (one for each 'bin'), there were many more descriptors (300+) than there were materials. Feature extraction was therefore performed by applying PLS regression. Leave-one-out cross-validation was used to select the number of extracted descriptors that would be used (resulting in one and two descriptors for, respectively, the prediction of adsorption and retention). An examination of the contribution of the original descriptors to the model showed the properties that were most influential to fibrinogen adsorption.

An application of these methods to high-throughput data was shown by Urquhart and co-workers (46). In this study, TOF-SIMS mass spectrometry measurements were obtained for 576 copolymers synthesized on an array. Using PCA, copolymer groups could be distinguished in the data based on their monomer constituents, and these groups could be related to the hydrocarbons measured by mass-spectrometry. Furthermore, a PLS prediction model was trained, relating TOF-SIMS measurements to wettability. The authors also published an analysis of these results for different subsets of the library, to observe the sensitivity of the prediction quality to the number of samples and the sample distribution (47).

Similar techniques have been applied to predict endothelial cell growth (48), fibrinogen adsorption on plasma deposited tetraglyme (49), the physical properties of aspirin tablets (50) or the wettability of patterned indium tin oxide surfaces (51). An interesting discussion on the analysis of absorbed protein films, using unsupervised and supervised feature extraction, can be found in (52), along with numerous other references to related work.

7.2.2   Molecular descriptors: towards hybrid surrogate-theoretical models

Since the inception of the materiomics field, many ideas that had their origin in the chemical-compound focused field of quantitative structure–activity relationships (QSAR), or quantitative structure–property relationships (QSPR), have been translated to a materials context. The QSAR/QSPR field is focused on predicting the properties of chemical compounds from their molecular structure. Relevant developments in this field started almost 50 years ago, with the first application of a linear regression model in 1964 (53). Since then, there has been a strong focus on the development of molecular descriptors (54, 55), which can be calculated through software such as the Molecular Operating Environment (MOE) (56, 57), the CODESSA package (58) and DRAGON (59). Through this software, molecules can be characterized on the basis of theoretical computations. Surrogate models can then be used to fill in the remaining gap between these theoretical descriptors and macroscopic complex material functions such as cell attachment. An early application in the materiomics field, in which surrogate models were combined with 101 theoretically calculated molecular descriptions of polymers, was described by Smith et al (15). An ANN was trained to predict protein adsorption and cell growth for polymers, based on molecular descriptors obtained from the MOE. As a relatively low number of polymers was measured (45–48), measurement errors had a potential large impact on the results. Interestingly, a Monte Carlo strategy was used to take this effect into account, by varying measured values according to experimental uncertainty, and observing the effects on the results. A feature selection was performed, showing that for the theoretical descriptors, the number of hydrogen atoms in a molecule and the octanol/water partition coefficient were especially relevant. However, the method also required experimentally determined descriptors (contact angle and glass transition temperature) to obtain an acceptable prediction performance. A later study (60) in which the method was combined with a PLS-based approach (61) reported the ability to predict adsorption based solely on theoretically derived descriptors.

Another application of molecular descriptors was shown in Li et al. (62), where drug release kinetics were related to polymer molecular structures, using support vector regression. A genetic algorithm was used to perform feature selection. The use of molecular descriptors has not, however, been limited to predicting polymer properties. Landrum, Penzotti and Putta used clustering, decision tree, nearest neighbour, SVM and bagging techniques to predict the performance of 96 homogeneous catalysts, by predicting the molecular weight of polymers they produced (63). Whereas molecular descriptors were originally often based on 2D representations of the molecule, with the increase in computation power 3D molecular structure descriptions have come into reach. While 3D methods were originally developed in the QSAR field to predict docking of drugs with proteins (64), they have also found application in the materiomics field. Cruz et al. applied 3D descriptors to the prediction of the activity of metallocene catalysts in ethylene polymerization (65). Molecules in a training set were aligned to each other based on their 3D structure. For each molecule, various fields were calculated (electrostatic, steric, lowest unoccupied molecular orbital (LUMO) and local softness), after which a PLS regression was used to relate these fields to the catalyst activity. Using this method, an accurate prediction of catalyst activity could be made. Linati *et al.* used 3D descriptors to find simple linear relations between descriptors and activity of bioglasses (66). For protein adsorption prediction, the step towards 3D molecular descriptions was made in Gubskaya *et al.* (13) (Figure 7.2). This study is described in more detail as a classic experiment (Box 7.2).

Not all types of measurements are scalable to high-throughput approaches. Several studies have therefore spent considerable effort on the optimization of the machine learning settings, in order to make optimal use of the available data. Roy, Potter and Landau analysed the role of various ANN settings in predicting ballistic impact resistance of polymers (67). Similarly, Gonzalez-Carrasco and co-workers compared (also for ballistic impact resistance) techniques such as multilayer perceptron (MLP), SVM and ANN, while optimizing algorithm settings using a genetic algorithm approach (68). For this specific setting, MLP was shown to be the best approach.

Many other studies could be mentioned. For example, Fourches *et al.* built models of nanomaterial toxicity (69). Using molecular descriptors of the nanomaterial particles, classification models of toxicity were built using SVM and nearest-neighbour classifiers. Also, Fourches *et al.* showed that the tested nanomaterials consisted of three different clusters with similar biological effects. Ghosh *et al.* predicted, based on 1664 molecular descriptors, fibrinogen adsorption, cell attachment and proliferation (70). A thorough review of these and other publications can be found in (71).

7.2.3   Morphological descriptors: describing the local cell environment

Compared with the work on molecular descriptors, far less work has been performed on computational methods that capture the effects of material structure morphology or, in other words, surface topologies. Although numerous works had already shown the influence of surface topology (e.g. 72–74), Meredith *et al.* only recently developed a gradient-based screen that captured the effects of polymer surface features on biological cells (75). The applied surface roughness and the diameter of PCL-rich surface topology features gradient strongly affected the expression of alkaline phosphatase. No further modelling was performed.

In an earlier study by Anselme *et al.*, cell proliferation was modelled as a function of surface roughness, based on 180 samples, measuring five different surface morphologies (76). To characterize the surfaces, a parameter describing surface organization (fractal dimension (77)) was used. Results showed a lower proliferation on less organized surfaces. The relation between proliferation, time and surface organization was modelled using a (least-squares fitted) manually constructed surrogate model, in contrast to using automated machine learning algorithms. This work showed that with an increase in fractal dimension, the contact area from cell to surface decreased, suggesting that cell proliferation and adhesion were related.

An interesting study was reported by Bigerelle and Anselme which determined for a total of 30 materials the long-term adhesion and proliferation capacity of cells (78). These samples were created using three different alloys, six different surface morphologies and two different roughness amplitudes. The use of surface descriptors in this work was limited to these design parameters. As the proliferation of cells affected the adhesion measurements, a theoretical model was developed that captured this effect, and used to correct the measurements. An ANOVA test was used to test for material, surface morphology and roughness effects in the measurements. Subsequently, a post-hoc ANOVA test (Duncan's multiple range test) was used to statistically distinguish the responses of the different samples. This test indicated that surfaces obtained through machine-tooling strongly induced cell proliferation, compared with techniques such as polishing and electro-erosion. Performance differences for the various chemical material compositions were not found to be statistically significant. Next, proliferation and adhesion measurements were related to each other by a least-square model. Significance was determined using a bootstrap protocol. It was shown that proliferation and adhesion were correlated.

Whereas in Zapata *et al.* (79) the results of the osteoblast screen reported in (75) were expanded by further characterizing the surface as well as cell morphology, in Su *et al.* (80) an interesting new approach was developed to characterize these cells in their local surface environment. It was noted that both cell–cell and cell–surface feature distances could affect proliferation, but that global surface metrics are unable to capture this information for (inhomogeneous) surfaces. By comparing the distance distributions with each other, as well as with random distance distributions, it was shown that proliferation was strongly dependent on the distance to, and proliferative status of, neighbouring cells.

Another study that focused on morphological material characterization was reported by Groeber *et al.* (81) In this study, 3D properties of materials, namely grain size, grain shape, number of grain neighbours and orientation, were determined from 2D sections. Next, the correlations between these descriptors were determined. Besides applications in surrogate modelling, these types of characterizations can also be used in more theoretical approaches, as was shown by Qidwai, Lewis and Geltmacher (82), who determined the plastic flow (yield) of metals in response to loading. This was accomplished using a finite-element modelling (FEM) approach based on a grain structure determined from 2D sections. These types of simulations lead, however, to enormous amounts of data, making the analysis of relations between microstructure properties and yield a difficult one. They suggested that machine learning techniques will play an important role in this aspect. A comprehensive approach to this type of analysis has been reported in Fullwood *et al.*(83)

Compared with other materiomics areas (particularly polymers), the number of experimental high-throughput methodologies to vary surface topographies has been somewhat limited. Recently, however, Unadkat *et al.* reported on a device allowing the characterization of 2178 different designed topographies simultaneously (84). Each of these topographies was designed using an algorithm-based approach, in which primitives (circles, triangles, rectangles) were combined into surface features. The input parameters for this design algorithm could be directly used as surface descriptors. Among others, it was shown that a predictor could be learned that, based on these descriptors, was able to distinguish surfaces with a low or high cell proliferation.

Box 7.3 Classic experiment

Designing optimal calcium phosphate scaffold–cell combinations

Modelling the dynamics of tissue formation can provide additional insights into the biological response to the material as well as providing an *in silico* tool to design and optimize the material. Bohner *et al.* (85) summarized the main difficulties in defining the optimal material properties and also proposed a new strategy to tackle this multidisciplinary problem: an integrative approach in which mathematical modelling is used to explain a mechanism of biomaterial–cell interactions, combined with experimental research to provide data for the determination of the model parameters as well as the validation of the model. This process requires both a careful design and extensive characterization of the scaffold. Moreover, it is inherently an iterative process in which new experimental results can be fed to the model and thorough model analysis can lead to new research hypotheses. This iterative process was followed by Carlier *et al.* (86) during the development and implementation of an experimentally informed bioregulatory model of the effect of calcium ions released from calcium phosphate (CaP)-based biomaterials on the activity of mesenchymal stem cell (MSC)-driven ectopic bone formation (schematic overview shown in Figure 7.5 A, B).

Bone formation is a very complex physiological process, involving the participation of many different cell types and regulated by countless biochemical, physical and mechanical factors, including naturally occurring or synthetic biomaterials. For the latter, CaP-based scaffolds have proven to stimulate bone formation, but at present still result in a wide range of *in vivo* outcomes, which is partly related to the sub-optimal use and combination with MSCs. To optimize CaP scaffold selection and make their use in combination with cells more clinically relevant, this study uses an integrative approach in which mathematical modelling is combined with experimental research.

The amount of bone formation predicted by the mathematical model corresponds to the amount measured experimentally under similar conditions (87, 88). Moreover, the model is also able to predict qualitatively the experimentally observed impaired bone formation under conditions such as insufficient cell seeding and scaffold decalcification. A strategy was designed *in silico* to overcome the negative influence of a low initial cell density on the bone formation process. Finally, the model was applied to design optimal combinations of calcium-based biomaterials and cell culture conditions with the aim of maximizing the amount of bone formation (Figure 7.5C). As with other models attempting to capture physiological processes, the determination of parameter values is a critical point in the setup of the model and defines the context in which the model can be used afterwards (qualitative versus quantitative). Carlier *et al.* (86) used a combination of experimentally derived and literature-based parameter values. The latter are generally adopted from other models and/or determined by means of a variety of experiments using *in vitro* and/or *in vivo* setups of different cell sources and species. In order to assess the importance of these parameters on the model outcome, a thorough sensitivity analysis needs to be performed.

Sensitivity analyses appear under many different forms. The most frequently used technique is the one-at-a-time (OAT) analysis where only one parameter is altered (e.g. 89). It provides information on the main effects of a parameter, but not on the combined effects or the interactions between different parameters. Carlier *et al.* (86) and others (90) have successfully applied design-of-experiment (DOE) techniques to overcome the limitations of the OAT technique.

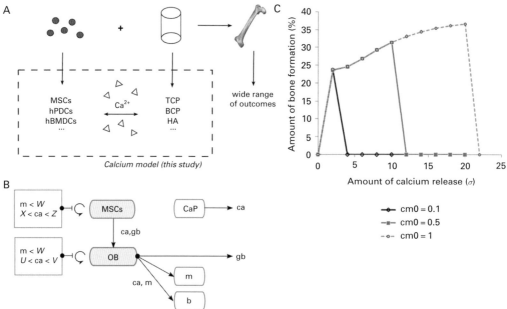

Figure 7.5   Design of optimal combinations of calcium phosphate scaffolds and cells, using an integrative model-based approach. A, Conceptual overview of the general bone tissue engineering process targeted in the model, combining stem cells (periosteal-derived, bone-marrow-derived or others) with a (calcium-releasing) carrier structure. hPDC, human periosteal derived stem cell; hBMDC, human bone marrow derived stromal cell; TCP, tri-calcium phosphate; BCP, bi-calcium phosphate; HA, hydroxy apatite. B, Schematic overview of the calcium model. W = maximum tissue density for proliferation, X = minimum calcium concentration for proliferation of MSCs, Z = maximum calcium concentration for proliferation of MSCs, U = minimum calcium concentration for proliferation of osteoblasts, V = maximum calcium concentration for proliferation of osteoblasts, ca = Ca^{2+}, gb = growth factor, m = osteoid, b = mineral matrix. The participation of a variable in a subprocess is indicated by showing the name of that variable next to the arrow representing that subprocess, e.g. calcium modulates differentiation and bone formation. C, Amount of bone formation at day 90 *in vivo* as a function of the calcium release rate (σ) and initial MSC density (cm0, in 10^6 cells/ml) as predicted by the mathematical model. Adapted with permission from (86).

7.2.4   Describing dynamic behaviour

Measuring material response under dynamic conditions is not always easy. Such dynamic conditions can range from changing biological processes to varying mechanical loads. These kinds of problems are often approached through the use of theoretical models. The field is far too large to discuss here in detail, so we will only give a few examples.

One of the interesting application areas of dynamic modelling techniques is in the modelling of living tissues. Some major focal points of this field have been morphogenesis (91) and models that describe the role of angiogenesis in tumour development (92). An example of an area in which both dynamic biological conditions and varying mechanical conditions come together is bone fracture healing. It has been shown that (limited) dynamic mechanical load can stimulate the fracture healing (93). Simultaneously, excessive load will disrupt the healing process. Determining exactly how this dynamic load influences the healing process is difficult. Geris, Sloten and Oosterwyck (94) therefore developed a theoretical mechanobioregulatory model of the healing process of a (needle fixated) murine tibial fracture. This model coupled the effect of a simulation of mechanical behaviour to an existing model of the bioregulatory processes in bone fracture healing. The modelling efforts showed that the mechanical load had to affect both angiogenesis and osteogenesis, to be able to describe healing failure under overload conditions. More recently, the same group developed a model that allows for a quantitative description of the biological processes, which is described in a classic experiment.

While the modelling of living tissues inevitably requires significant simplifications and assumptions, fully atomistic models of more complicated non-living biological materials are now coming within reach, as shown in a series of recent studies on the properties of spider silk and spider webs (95, 96, 18). Spider silk can achieve strengths which exceed those of steel, while using only weak H-bonds for chemical binding. To investigate how this is achieved, first the 3D conformation of proteins that make up the majority of the core of spider silk (MaSp1 and MaSp2) were investigated using MD simulations (103). Next, a mechanical analysis was performed, in which the structures were put under load. The relation between force and the stretching of the silk indicated the existence of three regimes: first, an initial stiff regime needed to be overcome, after which the spider silk yielded. Then a second very stiff regime was encountered, leading finally to failure of the silk. The models showed that this can be linked to the structure of the proteins: after the first stiff regime, the H-bonds in the amorphous domain start to break, leading to its stretching. When this domain is fully stretched, the multiple H-bonds linking the beta-sheets together cause the second stiff regime. Only when these beta-sheet crystals come apart does failure occur.

It was not yet clear, however, how these non-linear, multiple-regime characteristics contributed to the specific properties of the spider web. This was subsequently investigated by Cranford *et al.*(18) By modelling the web according to the atomistically derived loading responses, and comparing these with alternative loading responses,

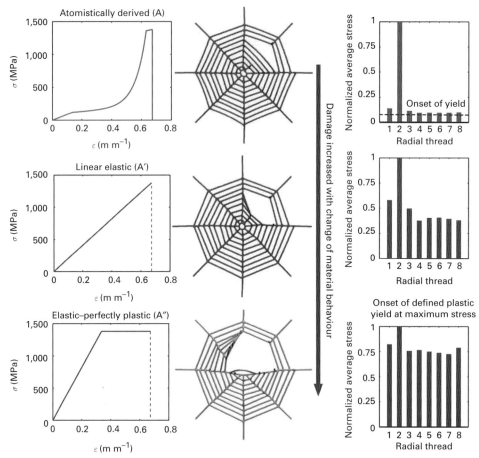

Figure 7.6   Three different models for the plasticity of spider silk, relating yield (x-axis) to load (y-axis). The relation to web failure and load distribution across the web is shown in the second and third columns. Reproduced with permission from (18).

they showed that the specific shape of the response curve ensures that, when the web is loaded till failure at a single position, the damage remains localized (Figure 7.6).

In terms of modelling techniques, an important aspect of this work is the connection it establishes between the atomistic nanometre scale behaviour and the behaviour at the scale of the spider web. To accomplish this, it uses a multiscale model architecture, in which a complex low-level model (of the protein structure) is summarized in a simpler model, fast enough to be able to simulate larger structures (the web). Such multiscale models have seen a lot of attention in the past two decades (97–100), as they enable a relatively complete understanding of the whole system, while keeping the simulation effort manageable. The simplified higher-scale models not only reduce computational effort but also allow one to obtain insight into individual roles of the specific layers of the system.

7.3   Future perspectives

Materials systems are complex: linked properties such as shape, surface topology, chemical compound structure and mechanical loading all play a role in affecting the material environment, which itself can consist of something as intricate as living tissue. To investigate and optimize these systems, the trend in the past 15 years has been to move from low-throughput experiments to high-throughput screening, allowing one to observe the combinatorial effect of the various material properties. Using these results, most studies perform some (limited) statistical inference.

The logical next step in understanding these material systems is the construction of models. This is a step that many studies do not yet take. It requires an interdisciplinary approach, combining material testing and computational approaches, in what could be called the 'materiomatics' field. As discussed in this chapter, such model construction can either be addressed in a data-driven way (constructing surrogate models), or in a hypothesis-driven way (constructing theoretical models). The difficulty in constructing theoretical models is that, in many cases, the complexity of material systems, in combination with our limited knowledge of the active processes, allows for such an enormous space of possible hypotheses that it prevents a rational approach to model construction. Surrogate models also have drawbacks, in particular that they often lack access to the researcher's implicit biological/material knowledge that is put to use in the construction of theoretical models.

One of the current trends is that researchers are attempting to combine the advantages of both modelling approaches, by incorporating more biological and material knowledge into their surrogate models. They do so by creating improved descriptors, based on hypotheses and/or modelling. A clear case of this is the molecular descriptors, where more and more theoretically derived properties are being used. It is likely that this trend will continue, with more and more advanced descriptors being developed. For example, in the context of surface descriptors, some of the first steps in this direction have already been made, in the form of cell centric surface descriptors (80). The culmination of this effort might be a fully theoretical model.

Another area that is likely to develop is the way in which high-throughput screens are performed. Currently, most studies perform a high-throughput screen followed by an analysis. Compared with this 'shotgun sampling' approach, more 'constructive' experimental designs could significantly increase the power of a study. Results obtained from previous experiments can be used to determine the maximally informative samples to test in subsequent iterations. The adoption of such iterative experimental designs will require new approaches to statistical inference and model building. Such methods can also benefit significantly from information obtained from experiments performed by other labs, but this will require significant efforts in standardization.

7.4   Snapshot summary

- Understanding how the complex interplay between material properties affects the performance of the whole material system requires not only the adoption of high-throughput screenings, but also the use of computational methods to model these interactions.
- Replicates are important. With more materials tested, more replicates per material are needed to correct for multiple testing effects.
- In selecting a material hit, it is important to take into account not only the observed performance, but also the variability of the replicate measurements, in order to assess one's confidence in a future high performance.
- Models allow one to determine to what extent experimental factors, such as material composition, are explanatory for material performance. Models based on theoretical knowledge are important as they allow one to falsify existing assumptions and/or suggest new explanations. When such theoretical knowledge is lacking, an *ab initio* modelling method (surrogate model) is a better choice. Many studies combine both approaches, by putting theoretical knowledge into their material descriptors, while using a surrogate model to capture how these descriptors can explain material performance.
- Models need to be validated, using test examples that have not been used to train the model. Cross-validation is a scheme that allows one to make efficient use of experimental results, by making it possible to use a large part of the dataset for model learning.
- In the absence of reliable knowledge on expected material performance, a good experimental design performs a uniform sampling of the material parameter space. If these parameters cannot be freely sampled, diversity methods are good alternatives.
- If not all experiments are performed in a batch, but rather sequentially, the amount of information gained per experimental sample can be significantly increased by balancing exploration and investigation of the material parameter space, through the use of active learning and optimization methods (experimental design).
- Numerous material system descriptors have been developed, ranging from molecular descriptors to the surface topology in a cell's neighbourhood. By putting theoretical knowledge into these descriptors, modelling and interpretation efforts can be helped significantly.

Further reading

Cranford S, Buehler M. Materiomics: biological protein materials, from nano to macro. Nanotechnol Sci Appl. 2010;**3**:127–48.

Fourches D, Pu D, Tassa C *et al.* Quantitative nanostructure–activity relationship modeling. ACS Nano. 2010;**4**(10):5703–12.

Kholodovych V, Smith J, Knight D *et al.* Accurate predictions of cellular response using QSPR: a feasibility test of rational design of polymeric biomaterials. Polymer. 2004;**45**(22):7367–79.

Kubinyi H. From narcosis to hyperspace: the history of QSAR. Quant Struct Activ Relat. 2002;**21**(4):348–56.

Neuss S, Apel C, Buttler P *et al*. Assessment of stem cell/biomaterial combinations for stem cell-based tissue engineering. Biomaterials. 2008;**29**(3):302–13.

References

1. Tabachnick B, Fidell L, Osterlind S. *Using Multivariate Statistics* 4th edn. Allyn & Bacon; 2001.
2. Noble W. How does multiple testing correction work? Nat Biotechnol. 2009;**27**(12):1135–7.
3. Benjamini Y, Hochberg Y. Controlling the false discovery rate: a practical and powerful approach to multiple testing. J Roy Stat Soc B.1995;**57**:289–300.
4. Swisher J, Jacobson S, Yücesan E. Discrete-event simulation optimization using ranking, selection, and multiple comparison procedures: A survey. ACM Trans Modeling Comput Simul (TOMACS). 2003;**13**(2):134–54.
5. Kim S, Nelson B. Selecting the best system. In *Handbooks in Operations Research and Management Science Vol. 13: Simulation*. North Holland; 2006. pp. 501–34.
6. Watanabe H, Khera S, Vargas M, Qian F. Fracture toughness comparison of six resin composites. Dental Mater. 2008;**24**(3):418–25.
7. Ashby M. Multi-objective optimization in material design and selection. Acta Mater. 2000;**48**(1):359–69.
8. Mutihac L, Mutihac R. Mining in chemometrics. Analyt Chim Acta. 2008;**612**(1):1–18.
9. Kardamakis A, Mouchtaris A, Pasadakis N. Linear predictive spectral coding and independent component analysis in identifying gasoline constituents using infrared spectroscopy. Chemomet Intell Lab Syst. 2007;**89**(1):51–8.
10. Jain A, Murty M, Flynn P. Data clustering: a review. ACM Comput Surveys (CSUR). 1999;**31**(3):264–323.
11. Wilkinson L, Friendly M. The history of the cluster heat map. Am Statist. 2009;**63**(2):179–84.
12. Handl J, Knowles J, Kell D. Computational cluster validation in post-genomic data analysis. Bioinformatics. 2005;**21**(15):3201–12.
13. Gubskaya A, Kholodovych V, Knight D, Kohn J, Welsh W. Prediction of fibrinogen adsorption for biodegradable polymers: Integration of molecular dynamics and surrogate modelling. Polymer. 2007;**48**(19):5788–801.
14. Mohamadi F, Richards N, Guida W *et al*. MacroModel? an integrated software system for modeling organic and bioorganic molecules using molecular mechanics. J Comput Chem. 1990; **11**(4): 440–67.
15. Smith J, Seyda A, Weber N *et al*. Integration of combinatorial synthesis, rapid screening, and computational modeling in biomaterials development. Macromol Rapid Commun. 2004;**25**(1):127–40.
16. Goddard W. A perspective of materials modelling. *Handbook of Materials Modeling*. Springer; 2005. pp. 2707–11.
17. Cranford S, Buehler M. Materiomics: biological protein materials, from nano to macro. Nanotechnol Sci Appl. 2010;**3**:127–48.
18. Cranford S, Tarakanova A, Pugno N, Buehler M. Nonlinear material behaviour of spider silk yields robust webs. Nature. 2012;**482**(7383):72–6.

19. Wold S, Hellberg S, Dunn IIIW. Computer methods for the assessment of toxicity. Acta Pharmacol Toxicol.1983;**52**:158–89.

20. Alpaydin E. *Introduction to Machine Learning*. MIT; 2004.

21. Byvatov E, Fechner U, Sadowski J, Schneider G. Comparison of support vector machine and artificial neural network systems for drug/nondrug classification. J Chem Inf Comput Sci. 2003;**43**(6):1882–9.

22. Friedman J, Hastie T, Tibshirani R. *The Elements Of Statistical Learning* Vol. 1. Springer; 2001.

23. Zweig M, Campbell G. Receiver-operating characteristic (ROC) plots: a fundamental evaluation tool in clinical medicine. Clin Chem. 1993;**39**(4):561–77.

24. Mason S, Graham N. Areas beneath the relative operating characteristics (ROC) and relative operating levels (ROL) curves: Statistical significance and interpretation. Quart J Roy Met Soc. 2002;**128**(584):2145–66.

25. Livingstone D, Salt D. Variable selection? Spoilt for choice? Rev Comput Chem. 2005: 287–348.

26. Guyon I, Elissee A. An introduction to variable and feature selection. J Machine Learning Res. 2003;**3**:1157–82.

27. Wessels L, Reinders M, Hart A *et al.* A protocol for building and evaluating predictors of disease state based on microarray data. Bioinformatics. 2005;**21**(19):3755–62.

28. Marée A, Grieneisen V, Hogeweg P. The cellular potts model and biophysical properties of cells, tissues and morphogenesis. In Anderson A, Rejniak K, eds. *Single-Cell-Based Models in Biology and Medicine*. Birkhäuser; 2007. pp. 107–36.

29. Cickovski T, Huang C, Chaturvedi R *et al.* A framework for three-dimensional simulation of morphogenesis. IEEE/ACM TransComput Biol Bioinformat. 2005;**2**(4):273–88.

30. Rejniak K, Anderson A. Hybrid models of tumor growth. WIRes Syst Biol Med. 2011;**3**(1):115–25.

31. Stéphanou A, McDougall S, Anderson A, Chaplain M. Mathematical modeling of the influence of blood rheological properties upon adaptative tumour-induced angiogenesis. Math Comput Modeling. 2006;**44**(1):96–123.

32. Myers R, Montgomery D, Anderson-Cook C. *Response Surface Methodology: Process and Product Optimization using Designed Experiments* Vol. 705. Wiley; 2009.

33. Yeten B, Castellini A, Guyaguler B, Chen W. A comparison study on experimental design and response surface methodologies. SPE Reservoir Simulation Symposium. Society of Petroleum Engineers Inc.; 2005. Available at http://www.onepetro.org/mslib/app/Preview.do?paperNumber =00093347&societyCode=SPE

34. Desai K, Survase S, Saudagar P, Lele S, Singhal R. Comparison of artificial neural network (ANN) and response surface methodology (RSM) in fermentation media optimization: Case study of fermentative production of scleroglucan. Biochem Eng J. 2008;**41**(3):266–73.

35. Harmon L. Experiment planning for combinatorial materials discovery. J Mater Sci. 2003;**38** (22):4479–85.

36. Hong L, Nelson B. A brief introduction to optimization via simulation. *Proc 2009 Winter Simulation Conf (WSC)*. IEEE; 2009. pp. 75–85.

37. Settles B. Active learning literature survey. *Computer Sciences Technical Report 1648*. University of Wisconsin-Madison; 2009.

38. Adams N, Murray-Rust P. Engineering polymer informatics: Towards the computer-aided design of polymers. Macromol Rapid Commun. 2008;**29**(8):615–32.

39. Berners-Lee T, Hendler J. Scientific publishing on the semantic web. Nature. 2001;**410**:1023–4.

40. Blake J, Bult C. Beyond the data deluge: data integration and bio-ontologies. J Biomed Informat. 2006;**39**(3):314–20.

41. Bodenreider O, Stevens R. Bio-ontologies: current trends and future directions. Briefings Bioinformat. 2006;**7**(3):256–74.

42. Xiang X, Sun X, Briceno G *et al.* A combinatorial approach to materials discovery. Science. 1995;**268**(5218):1738.

43. Danielson E, Golden J, McFarland E *et al.* A combinatorial approach to the discovery and optimization of luminescent materials. Nature. 1997;**389**(6654):944–8.

44. Jandeleit B, Schaefer D, Powers T, Turner H, Weinberg W. Combinatorial materials science and catalysis. Angew Chem Int Ed. 1999;**38**(17):2494–532.

45. Pérez-Luna V, Horbett T, Ratner B. Developing correlations between fibrinogen adsorption and surface properties using multivariate statistics. J Biomed Mater Res. 1994;**28**(10):1111–26.

46. Urquhart A, Taylor M, Anderson D *et al.* ToF-SIMS analysis of a 576 micropatterned copolymer array to reveal surface moieties that control wettability. Analyt Chem. 2008;**80** (1):135–42.

47. Taylor M, Urquhart A, Anderson D *et al.* Partial least squares regression as a powerful tool for investigating large combinatorial polymer libraries. Surf Interface Anal. 2009;**41**(2):127–35.

48. Chilkoti A, Schmierer A, Pérez-Luna V, Ratner B. Relationship between surface chemistry and endothelial cell growth: Partial least-squares regression of the static secondary ion mass spectra of oxygen-containing plasma-deposited films. Analyt Chem. 1995;**67** (17):2883–91.

49. Shen M, Wagner M, Castner D, Ratner B, Horbett T. Multivariate surface analysis of plasma-deposited tetraglyme for reduction of protein adsorption and monocyte adhesion. Langmuir. 2003;**19**(5):1692–9.

50. Taylor M, Elhissi A. Predicting the physical properties of tablets from atr-ftir spectra using partial least squares regression. Pharm Devel Technol. 2011;**16**(2):110–7.

51. Yang L, Shard A, Lee J, Ray S. Predicting the wettability of patterned ito surface using tof-sims images. Surf Interface Anal. 2010;**42**(6–7):911–15.

52. Wagner M, Tyler B, Castner D. Interpretation of static time-of-flight secondary ion mass spectra of adsorbed protein films by multivariate pattern recognition. Analyt Chem. 2002;**74** (8):1824–35.

53. Kubinyi H. From narcosis to hyperspace: the history of QSAR. Quant Struct–Activity Relat. 2002;**21**(4):348–56.

54. Todeschini R, Consonni V. *Handbook of Molecular Descriptors* Vol. 79. Wiley; 2008.

55. Tseng Y, Hopfinger A, Esposito E. The great descriptor melting pot: mixing descriptors for the common good of qsar models. J Computer-aided Mol Design. 2012;**26**:39–43.

56. Chemical Computing Group Inc. Montreal Q. Molecular operating environment. 2012. http://www.chemcomp.com.

57. Vilar S, Cozza G, Moro S. Medicinal chemistry and the molecular operating environment (MOE): application of QSAR and molecular docking to drug discovery. Curr Topics Med Chem. 2008;**8**(18):1555–72.

58. Karelson M, Maran U, Wang Y, Katritzky A. QSPR and QSAR models derived using large molecular descriptor spaces. a review of codessa applications. Collection of Czechoslovak Chem Commun. 1999;**64**(10):1551–71.

59. Talete. DRAGON. 2012. http://www.talete.mi.it/products/dragon_description.htm.

60. Smith J, Kholodovych V, Knight D, Kohn J, Welsh W. Predicting fibrinogen adsorption to polymeric surfaces in silico: a combined method approach. Polymer. 2005;**46** (12):4296–306.
61. Kholodovych V, Smith J, Knight D *et al*. Accurate predictions of cellular response using qspr: a feasibility test of rational design of polymeric biomaterials. Polymer. 2004;**45** (22):7367–79.
62. Li X, Petersen L, Broderick S, Narasimhan B, Rajan K. Identifying factors controlling protein release from combinatorial biomaterial libraries via hybrid data mining methods. ACS Combinat Sci. 2011;**13**(1):50–8.
63. Landrum G, Penzotti J, Putta S. Machine-learning models for combinatorial catalyst discovery. Measurement Sci Technol. 2005;**16**:270.
64. Lyne P. Structure-based virtual screening: an overview. Drug Discovery Today. 2002;**7** (20):1047–55.
65. Cruz V, Ramos J, Martinez S *et al*. Structure–activity relationship study of the metallocene catalyst activity in ethylene polymerization. Organometallics. 2005;**24**(21):5095–102.
66. Linati L, Lusvardi G, Malavasi G *et al*. Qualitative and quantitative structure-property relation-ships analysis of multicomponent potential bioglasses. J Phys Chem B. 2005;**109**(11):4989–98.
67. Roy N, Potter W, Landau D. Polymer property prediction and optimization using neural networks. IEEE Trans Neural Netw. 2006;**17**(4):1001–14.
68. Gonzalez-Carrasco I, Garcia-Crespo A, Ruiz-Mezcua B, Lopez-Cuadrado J. An optimization methodology for machine learning strategies and regression problems in ballistic impact scenarios. Appl Intell. 2010:**36**:1–18.
69. Fourches D, Pu D, Tassa C *et al*.Quantitative nanostructure- activity relationship (QNAR) modelling. ACS Nano. 2010;**4**(10): 5703–12.
70. Ghosh J, Lewitus D, Chandra P *et al* Computational modeling of in vitro biological responses on polymethacrylate surfaces. Polymer. 2011;**52**(12):2650–60.
71. Tu Le, Epa V, Burden F, Winkler D. Quantitative structure–property relationship modeling of diverse materials properties. Chem Rev. 2012;**112**(5):2889–919.
72. Lampin M, Warocquier-Clérout R, Legris C, Degrange M, Sigot-Luizard M. Correlation between substratum roughness and wettability, cell adhesion, and cell migration. J Biomed Mater Res. 1997;**36**(1):99–108.
73. Miller C, Shanks H, Witt A, Rutkowski G, Mallapragada S. Oriented schwann cell growth on micropatterned biodegradable polymer substrates. Biomaterials. 2001;**22**(11):1263–9.
74. Hatano K, Inoue H, Kojo T *et al*. Effect of surface roughness on proliferation and alkaline phosphatase expression of rat calvarial cells cultured on polystyrene. Bone. 1999;**25**(4):439–45.
75. Meredith J, Sormana J, Keselowsky B *et al*. Combinatorial characterization of cell interactions with polymer surfaces. J Biomed Mater Res A. 2003;**66**(3):483–90.
76. Anselme K, Bigerelle M, Noel B *et al*. Qualitative and quantitative study of human osteoblast adhesion on materials with various surface roughnesses. J Biomed Mater Res. 2000;**49** (2):155–66.
77. Benhamou C, Lespessailles E, Jacquet G *et al*. Fractal organization of trabecular bone images on calcaneus radiographs. J Bone Miner Res. 1994;**9**(12):1909–18.
78. Bigerelle M, Anselme K. Statistical correlation between cell adhesion and proliferation on biocompatible metallic materials. J Biomed Mater Res Part A. 2005;**72**(1): 36–46.
79. Zapata P, Su J, García A, Meredith J. Quantitative high-throughput screening of osteoblast attachment, spreading, and proliferation on demixed polymer blend micropatterns. Biomacromolecules. 2007;**8**(6):1907–17.

80. Su J, Zapata P, Chen C, Meredith J. Local cell metrics: a novel method for analysis of cell-cell interactions. BMC Bioinformat. 2009;**10**(1):350.

81. Groeber M, Ghosh S, Uchic M, Dimiduk D. A framework for automated analysis and simulation of 3D polycrystalline microstructures. Part 1: Statistical characterization. Acta Mater. 2008;**56**(6):1257–73.

82. Qidwai M, Lewis A, Geltmacher A. Using image-based computational modeling to study microstructure-yield correlations in metals. Acta Mater. 2009;**57**(14):4233–47.

83. Fullwood D, Niezgoda S, Adams B, Kalidindi S. Microstructure sensitive design for performance optimization. Prog Mater Sci. 2010;**55**(6):477–562.

84. Unadkat H, Hulsman M, Cornelissen K *et al*. An algorithm-based topographical biomaterials library to instruct cell fate. Proc Natl Acad Sci. 2011;**108**(40):16565–70.

85. Bohner M, Loosli Y, Baroud G, Lacroix D. Commentary: deciphering the link between architecture and biological response of a bone graft substitute. Acta Biomater. 2011;**7**(2):478–84.

86. Carlier A, Chai Y, Moesen M *et al*. Designing optimal calcium phosphate scaffold-cell combinations using an integrative model-based approach. Acta Biomater. 2011;**7**(10):3573–85.

87. Hartman E, Vehof J, Spauwen P, Jansen J. Ectopic bone formation in rats: the importance of the carrier. Biomaterials. 2005;**26**(14):1829–35.

88. Roberts S, Geris L, Kerckhofs G *et al*. The combined bone forming capacity of human periosteal derived cells and calcium phosphates. Biomaterials. 2011;**32**(19):4393–405.

89. Geris L, Gerisch A, Maes C *et al*. Mathematical modeling of fracture healing in mice: comparison between experimental data and numerical simulation results. Med Biol Eng Comput. 2006;**44**(4):280–9.

90. Isaksson H, van Donkelaar C, Huiskes R, Yao J, Ito K. Determining the most important cellular characteristics for fracture healing using design of experiments methods. J Theor Biol. 2008;**255**(1):26–39.

91. Urdy S. On the evolution of morphogenetic models: mechano-chemical interactions and an integrated view of cell differentiation, growth, pattern formation and morphogenesis. Biol Rev. 2012;**87**(4):786–803.

92. Jiang Y, Bauer A, Jackson T. Cell-based models of tumor angiogenesis. In *Modeling Tumor Vasculature*. Springer; 2012. pp. 135–50.

93. Kenwright J, Gardner T. Mechanical influences on tibial fracture healing. Clin Orthopaed Related Res. 1998;**355**:S179.

94. Geris L, Sloten J, Oosterwyck H. Connecting biology and mechanics in fracture healing: an integrated mathematical modeling framework for the study of nonunions. Biomech Modeling Mechanobiol. 2010;**9**(6):713–24.

95. Keten S, Buehler M. Nanostructure and molecular mechanics of spider dragline silk protein assemblies. J Roy Soc Interface. 2010;**7**(53):1709–21.

96. Nova A, Keten S, Pugno N, Redaelli A, Buehler M. Molecular and nanostructural mechanisms of deformation, strength and toughness of spider silk fibrils. Nano Lett. 2010;**10**(7):2626–34.

97. Lépinoux J. *Multiscale Phenomena in Plasticity: From Experiments to Phenomenology, Modeling and Materials Engineering* Vol. 367. Springer; 2000.

98. Ghoniem N, Cho K. The emerging role of multiscale modeling in nano-and micro-mechanics of materials. Comput Modeling Eng Sci. 2002;**3**(2):147–74.

99. Curtin W, Miller R. Atomistic/continuum coupling in computational materials science. Modeling Simul Mater Eci Eng. 2003;**11**:R33.

100. Phillips R. Multiscale modeling in the mechanics of materials. Curr Opin Solid State Mater Sci. 1998;**3**(6):526–32.

101. Bock H. On some significance tests in cluster analysis. J Classif. 1985;**2**(1):77–108.

102. Čopíková J, Barros A, Šmídová I *et al*. Influence of hydration of food additive polysaccharides on FT-IR spectra distinction. Carbohyd Polym. 2006;**63**(3):355–59.

103. Keten S, Xu Z, Ihle B, Buehler M. Nanoconfinement controls stiffness, strength and mechanical toughness of [beta]-sheet crystals in silk. Nat Mater. 2010;**9**(4):359–67.

8 Upscaling of high-throughput material platforms in two and three dimensions

Gustavo A. Higuera, Roman K. Truckenmüller, Rong Zhang,
Salvatore Pernagallo, Fabien Guillemot and Lorenzo Moroni

Scope

High-throughput screening (HTS) is carried out on two- (2D) and three-dimensional (3D) materials, with hundreds to thousands of conditions at various size scales. When hits are successfully found in HTS systems, upscaling to clinically relevant surfaces needs to be performed to validate whether the identified material and functionality can be replicated on the macroscale. In doing so, parameters such as surface chemistry, topography and sample dispensing must be controlled to maintain reproducibility. Here, we discuss the methods harnessed to replicate chemical and topographical features from the nano- to the macroscale in 2D and 3D systems. Technologies to control cell adhesion and 3D scaffold fabrication are introduced and discussed in terms of their potential for HTS.

8.1 Basic upscaling principles

HTS is a highly automated process that tests small amounts of large numbers of compounds for a desired function. In the previous chapters of this book, the general principles behind material chemistry and resulting physico-chemical properties, combinatorial chemistry, microfabrication technologies and development of tools to perform biological assays on HTS platforms have been described. These elements partly return here, where basic principles of polymer chemistry and surface topographies are introduced in the context of facing the technological challenges to upscale selected candidates to larger surfaces or medical devices with complex curved shapes. In addition, the basic principles behind implant fabrication technologies and precise cell deposition are discussed to illustrate the steps required to assimilate HTS into clinically relevant 3D systems.

Materiomics: High-Throughput Screening of Biomaterial Properties, ed. Jan de Boer and Clemens van Blitterswijk. Published by Cambridge University Press. © Cambridge University Press 2013.

8.1.1   Basic processes for the generation of chemical libraries

Materials identified in a first HTS ('hit' candidates) are re-synthesized in upscaled amounts (from pico- or nanograms up to grams) to allow validation of results that emerged from the initial screening. This is a necessary step in a modern high-throughput material development cycle. It allows validation of the results obtained from the initial material screening via (i) the re-synthesis of the substrates being screened; and (ii) the generation of more detailed data, such as full characterization of cells harvested from large culture substrates.

The basic principle of upscaling candidates is to replicate them on the macroscale using the same synthesis road used for the library preparation. Replication provides the information required to give confidence in the ability to re-synthesize these 'hits' and to use the same synthetic procedures and reagents to produce analogies in follow-up work. The chemical and physical properties of the identified candidates, prepared on the large scale, have to be analysed and compared with their counterparts from libraries at the small scale to make sure that they are the same or close enough for further study. Inappropriate follow-up work may lead to 'false positives' because of batch-to-batch deviations of the synthesis process, or differences in reagent purity and modes of operation (1).

In the primary HTS, compounds are only tested in duplicate with volumes ranging from pico- up to microlitres. For some screening targets (e.g. cells), a large number of 'hit' candidates could turn up in the primary screening. If positive results or 'hits' are discovered in a primary screen, a more accurate and precise secondary screening is performed by preparing and carrying out a subsequent 'hit' library for screening (2–4). It is unrealistic to scale up all the candidates, owing to limited experimental resources. Thus, a secondary screening with more replicates for an individual candidate can effectively narrow down the selection of candidates that were found positive in the primary screening.

The secondary screening does not need to be under the same condition as the initial HTS. Generally, more informative conditions may be applied to screen the top candidates further. For example, polymer hydrogel microarrays with thousands of various polymers have been recently developed (Figure 8.1A and B(a)) and used to identify substrates for cellular trapping and thermally triggered release of an immortal cell line (HeLa GFP) (4). All polymer spots on the array were imaged after 48 hours of incubation at 37 °C, and subsequent incubations at 20 °C (Figure 8.1B). 'Hit' polymers from the primary screening were reprinted with multiple copies to obtain a 'hit-microarray' for a secondary screening and to confirm the polymer top candidates (Figure 8.1C). However, polymer spots on the 'hit-microarray' were still about 100 μm to 500 μm in diameter and therefore not fully informative for an eventual upscaled *in vitro* study. Thus, the hits that emerged were upscaled gradually in several steps. To do so, either 24- or 48-well plates were used to provide enough cells to study and elucidate cell–candidate interactions. A further screening used spin-coated or polymer-grafted cover slips in six well plates for the best candidates (Figure 8.1D). A couple of candidates were pinpointed for long-duration and multi-passage cell culture studies, and subsequently processed in industrial settings for preliminary tests before commercialization.

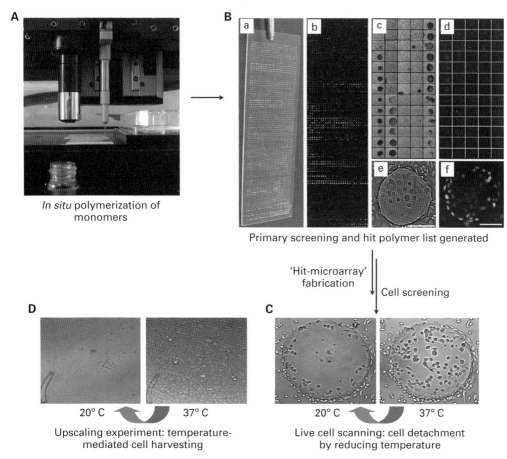

Figure 8.1 Upscaling thermo-sensitive hydrogels identified from polymer hydrogel microarray screening of HeLa GFP cells. A, Fabrication of microarrays with an inkjet printer through in situ polymerization of monomers on glass slides. B, Polymer microarrays with over 2000 polymers on one slide and their cellular screening and analysis: (a) a polymer hydrogel microarray, (b) microarray of a primary screening of HeLa cells treated with DAPI, (c) and (d) mosaic of bright-field images and fluorescent images after microarray scanning, (e) and (f) images of a typical hit hydrogel spot. C, Hit microarray development and screening. Polymer hydrogel microarrays were then scanned under different temperature to identify thermosensitive hydrogels for cell binding and temperature-triggered detachment (HeLa cells, 40 min incubation at 20 °C). D, Upscaling of identified candidates on coverslips for various cell culture and temperature-mediated detachment (L929 cells, 1 hour incubation at 20 °C). Scale bars: 200 μm.

8.1.2   Basic processes for the generation of micro- and nanotopographical patterns

For the generation of surface micro- and nanotopographies, a large number of patterning processes exist (5). They can be used to create HTS libraries of topographies enhancing the bioactivity of medical implants (see also Chapter 4). Topographical patterning processes can be classified as *subtractive* or *additive* processes, depending on whether the patterning process removes material from or adds material to a substrate, or as *formative* or moulding processes. Furthermore, patterning processes can also be classified as *serial* or *parallel* processes, in the case of lithography, for example, depending on whether a pattern is generated by a multiexposure, (i.e. step by step), or by a single exposure. Serial lithographic processes need no masks and can have high resolutions, but have low throughput. In contrast, parallel lithographic processes need masks and have much higher throughput, but lower resolution.

Processes for the generation of micro- and nanotopographies include – among others – lithography in radiation-sensitive polymer resists, based on focused light or particle beams, such as laser lithography or laser direct writing, and electron-beam (EBL) or ion-beam (IBL) lithography (6). Sometimes these processes are applied as direct lithography where the resist patterned on the substrate is maintained in the final application. More often, however, they are used to generate a temporary resist mask. This is then used for subsequent subtractive pattern transfer by wet or dry etching into the substrate, or additive pattern transfer by build-up of a material on the substrate via electroplating or vapour deposition. An alternative is to use mask-based lithography such as ultraviolet, deep ultraviolet (DUV) or X-ray lithography (6), with the masks again being fabricated by beam-based patterning processes. X-ray lithography is performed as 1:1 shadow projection. Ultraviolet or photolithography is not only performed as 1:1 contact or proximity printing, but also as projection exposure with a reduction lens between mask and substrate. In the latter case, in a 'stepper', the pattern of a 'reticle' is repeatedly copied in the photoresist.

Focused beams can also be used to pattern the substrate directly without the use of resists, particularly in laser ablation or with focused ion beams. The interference of coherent light waves in laser interference lithography (7, 8) or of coherent electron waves can be used for the maskless but parallel generation of regular, comparatively simple submicrometre or nanometre patterns, respectively.

Processes for micro- or nanotopography generation also include the moulding or replication processes called micro-injection moulding (9) and hot embossing (10). Micro-injection moulding is also applied to fabricate miniature biomedical parts such as hearing aid filters or bioresorbable haemostatic clips. Hot embossing, in the biomedical field mainly used to fabricate microfluidic biochips and cell culture microwell arrays, also includes soft embossing with soft elastomeric moulds cast from corresponding masters (11, 12). The so called 'soft lithography' processes also use such elastomeric moulds.(7) Micro- or nano-moulding typically enables high manufacturing throughput. In case of high patterning resolutions, the corresponding moulds are fabricated using lithography-based processes. Nanoimprint lithography (NIL) uses moulds instead of photomasks to pattern resists thermally or in combination with UV curing (13).

Micro-electrical discharge machining (14) and micro-electrochemical machining (15) also allow the serial or parallel patterning of conductive substrates according to the shape of an electrode pin or stamp. The electrode is patterned via means described above (hot embossing, soft lithography etc.).

8.1.3   Translation to three dimensions

Once material chemistry and surface topography candidates have been found from HTS systems, medical implants reflecting these newly discovered 'hits' can be fabricated. A number of conventional and novel technologies can be considered for this.

Moulding has been historically used to manufacture devices of diverse shape and composition, such as total joint prostheses (16, 17). Different moulding technologies are nowadays established, such as compression, transfer and injection moulding. All these technologies rely on inserting the selected raw material into a chamber with a desired shape – a mould. The final product is then obtained by applying heat and pressure. Similarly, textile technologies have been widely used to produce artificial vascular grafts, and more recently porous biomaterials – scaffolds – have been used as a vehicle for cells and biological factors in tissue engineering and regenerative medicine (18, 19).

Typically, textile meshes are created by spinning a biomaterial solution into a coagulation bath where they are collected on drums, subsequently washed and dried. Depending on the speed of the drums and on the affinity between the solvent of the biomaterial solution and the non-solvent of the coagulation bath, fibres with different dimensions and shape can be obtained. Finally, textile meshes are obtained by random deposition and eventually heat or chemical bonding (non-woven), or by weaving and knitting.

Undoubtedly the most promising route to translate candidates from HTS platforms to final clinical products currently encompasses the design of textile scaffolds with instructive properties for the regeneration of tissues and organs. Textile products are versatile and possess high tensile mechanical properties. Their fibres can be oriented to form meshes with tailored geometrical properties. Furthermore, they can be modified to provide various surface properties, including random topographies, depending on the processing conditions.

Foaming technologies have also been extensively used to create highly porous sponges. Foams are created by inserting the biomaterial of choice and a porogen agent into a mould. The system is then heated under pressure to allow melting or fusion of the raw biomaterial, cooled, and washed to remove the porogen agent. Alternatively, biomaterial solutions can be used, to avoid heating. In this case, the solvent is left to evaporate, before washing out the porogen agent. By controlling the size and shape of the particles used to create the foam, it is possible to customize its pore size, shape and total porosity (20). Alternatively, gases can be used to create a porous biomaterial where controlling the gas flow rate, pressure and type leads to variations in the resulting pore network (21).

A more recently established and flexible platform of technologies with the capacity to create versatile products is rapid prototyping (22–24). Despite differences in the methods of manufacturing, this set of techniques is united by the same work-flow principles, where a computer aided design (CAD) model is divided into a stack of slices and produced by computer aided manufacturing (CAM) in a layer-by-layer manner (Figure 8.2). In extrusion-based rapid prototyping (e.g. fused deposition modelling, direct and 3D printing), the biomaterial is either directly extruded from a syringe or

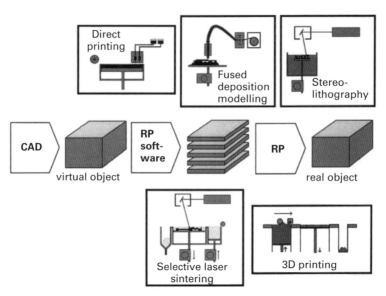

Figure 8.2   Schematic representations of some of the rapid prototyping (RP) technologies most used to fabricate porous biomaterials. Scaffolds are designed by CAD software to create a virtual object, which is transferred to the CAM software governing the rapid prototyping equipment and divided in a stack of slices or layers. By controlling the processing parameters of each specific technology, layers are produced on top of each other in a layer-by-layer fashion. Reprinted with permission from (23).

placed as a particle bed on a stage. In the former case, the biomaterial is extruded in the form of a fibre with the application of pressure at a predetermined temperature, which is selected to be either at room or physiological temperature (e.g. hydrogels, highly viscous biomaterial solutions/slurries) or above the biomaterial melting temperature. In the latter case, a binding agent is extruded on the biomaterial bed, thus allowing binding of different particles on its trajectory. In laser-based rapid prototyping, an implant is fabricated either by selective light exposure of liquid photosensible biomaterials (stereolithography), via the use of a mask and a projecting lens system, or by selectively moving a laser beam onto a biomaterial particle bed, thus causing particle binding locally along the laser trajectory (selective laser sintering). Layer-by-layer fabrication allows tailoring the pore network and producing patient-customized medical devices with anatomical shapes derived by computer tomography or magnetic resonance imaging datasets.

An alternative method to reproduce HTS surface topographies and combinatorial chemistry hits bypasses the conventional idea of imparting these hits by combination of different technologies and attempts to do so directly by varying chemical properties and microstructure during fabrication. Fabrication technologies that allow genuine control over constructed features at the micro- and nanoscales include electrospinning (25) and molecular self-assembly (26). In electrospinning, a biomaterial solution is injected into a high-voltage electrostatic field. This induces an unstable state in the solution, resulting in the formation of fibres which are collected on a target grounded to earth.

Depending on the applied flow rate, solution concentration, field strength and environmental conditions, textile meshes with fibre dimensions of a few micrometres to a few tens of nanometres can be obtained. Molecular self-assembly consists in the design of molecular building blocks with a chemical conformation that spontaneously promotes assembly of complementary units. These units can be tailored to present specific biological moieties such as peptide sequences and typically result in the formation of hydrogel nanofibrillar networks.

In the endeavour to create biomedical implants that can mimic the complexity of tissues and organs to be repaired in our body, all the fabrication technologies discussed above can be combined with other rapid prototyping systems that also allow precise control over spatial distribution of cells in 2D and 3D upscaled systems. These dispensing technologies are able to process cells alone or in combination with biocompatible materials (bio-inks). Current cell-dispensing technologies can be divided into three main categories: (i) inkjet printing, (ii) 3D plotting and (iii) laser-assisted bioprinting.

Since pioneering work by Klebe et $al.$ (27), commercial inkjet printers (e.g. piezoelectric and thermal) have been modified to print micro-droplets of biological materials, including viable cells, with microscale precision (28–30) at a printing speed of up to 10 kHz (31). Although this technology allows cell viability to be maintained (31, 32), 'as-fabricated' tissue constructs contain a low cell density (33), which is in contrast to the cellular volume fraction found in most living systems. This limitation comes from the need to use low cell concentrations (up to 5×10^6 cells ml^{-1}) (6, 34) in the bio-ink, to avoid clogging due to cell sedimentation and aggregation (31).

Bioplotting comprises extrusion-based rapid prototyping methods that have been developed to control the geometry of engineered tissues. A pneumatic dispensing system is used to print 3D structures by depositing a continuous stream of cell-laden hydrogel. A cellularized scaffold with geometric control of its internal structure and external shape is then created layer-by-layer (35). Moreover, 3D plotting can be used to print cell aggregates, used as building blocks, without hydrogel (36). While these methods offer a great opportunity to handle highly concentrated cell-containing media, their typical features have low resolution with minimum lateral dimensions of a few hundred micrometres.

Laser-assisted bioprinting (LAB) (37) is based on the principle of laser-induced forward transfer (LIFT) (38). A typical LAB setup comprises a pulsed laser beam, a focusing system, a 'ribbon' (i.e. a transparent glass slide, possibly coated by a laser-absorbing layer of metal, on which a thin layer of bio-ink is spread), and a receiving substrate facing the ribbon. Its physical principle is based on the generation of a cavitation-like bubble into the depth of the bio-ink film, whose expansion and collapse induce the formation of a jet, and thereby the transfer of the bio-ink from the ribbon to the substrate. In order to print cells with high resolution and high throughput, parameters related to laser pulse characteristics (wavelength, pulse duration, repetition rate, energy, beam focus diameter), bio-ink properties (viscosity, thickness, surface tension) and substrate characteristics should be adjusted. Previous studies have shown that LAB can print mammalian cells without affecting viability and function, and without causing DNA damage. LAB is a nozzle-free technology that prevents clogging issues and allows the printing of droplets from solutions of various viscosities (1–300 mPa s^{-1}) and with cell concentration of the order of 1×10^8 cells ml^{-1}.

Nevertheless, it currently suffers from several drawbacks. Most importantly, the metallic laser-absorbing layer is vaporized during the printing process, and metallic residues might be observed in the final product.

8.2   Upscaling and materiomics

Technology upscaling from HTS platform to biomedical implants and from 2D to 3D presents a unique set of challenges. Batch-to-batch variations and operation modes during biomaterial synthesis, accessibility to curved surfaces and shadow area, production speed, and multiple cell/material dispensing are some examples of parameters that need to be anticipated for upscaling. Even when this is properly done, there remains a persistent question: can we fully translate results obtained on 2D surfaces into 3D systems, or do we need to re-invent the wheel?

8.2.1   Upscaling of chemical libraries

High-throughput screening of chemical compounds to identify modulators of molecular targets is a mainstay of pharmaceutical development. HTS is being increasingly used to identify chemical probes of small molecules (1, 39, 40), inorganic chemicals (41, 42), peptides/proteins (43, 44), polymers (45) and many other chemical species (46). However, challenges still need to be surmounted during upscaling of the hit candidates.

Beside the purity control of the large-scale synthesis of some complex organic chemicals or peptides, another challenge is that the hit candidates are difficult to scale up for downstream optimization because they often are difficult, or too expensive, to synthesize on the large scale. Therefore, it is important to consider chemical availability during library preparation.

For cell-based biochemical HTS, the identified candidates such as peptides may face a much more complicated environment upon upscaling tests in tissue systems such as animal or human bodies. The main challenge is that the identified proteins or other biochemicals could interact not just with target cells but also with other biochemicals in the body that may alter their biological efficiency and hamper the specificity observed *in vitro*.

With biomaterials, especially synthetic polymers patterned in miniaturized 3D format, the characterization and analysis of their biological response are different from 2D substrates. The complexity of 3D biomaterial arrays is often responsible for the large deviations observed during the biological assessment when compared with 2D substrates. This makes comparison of results difficult and upscaling more uncertain. Hence, it could be crucial to develop and use advanced imaging approaches to obtain accurate information from 3D cell–material interactions. In addition, bringing forward other candidates, not picked out in the first screen but chemically related to those 'hits', for secondary or even tertiary screening under upscaled conditions might reduce the uncertainty that we currently observe when stepping from 2D surfaces to 3D constructs.

As mentioned above, the use of a high-throughput approach such as polymer microarrays to allow the rapid screening of chemically diverse polymers offers an important

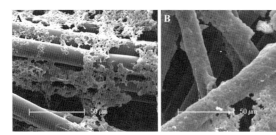

Figure 8.3   Cultivation of hESC-derived hepatocyte-like cells onto a polymer-coated scaffold. A, Polyurethane 134-coated scaffold; B, bio-artificial liver (BAL) devices. Scale bar: 50 μm.

tool to find correlations between the design and performance of such materials (47, 48). This may help, for instance, to understand cellular binding and growth on 'hit' polymers (30). Conditions applied for the library screening must echo the application conditions. Characterization of screening should address reliable protocols and techniques, in order to provide constant and comparable results. Recently, polymer libraries based on 1000 polyacrylates synthesized by free radical copolymerization using 30 diverse acrylate-monomers, and on 1000 polyurethanes synthesized by condensation polymerization of six diols, five di-isocyanates and nine chain extenders have been successfully upscaled to the gram scale (49–52). A polyurethane polymer (PU134) has been identified from the library screening for supporting growth of human embryonic stem cell (hESC)-derived hepatic endoderm. PU134 coated scaffolds (1 × 1 cm) were further analysed (50) and shown to improve cell growth (Figure 8.3).

Polymer microarrays prepared by *in situ* polymerization have also been developed by using an inkjet and a contact printing method (4, 50, 53), generating a hydrogel micro-array with over 2000 different polymers for cellular screening. In the fabrication of polymer microarrays, several replicates of the same polymer are used (3, 52). However, for primary screening of big libraries, repeating spots may be unnecessary (4), as is often also reported in drug discovery screening (1).

8.2.2   Upscaling of topographical libraries

When transferring 'hits' from HTS of miniaturized, planar, chip-type libraries of top-ographies to the surfaces of medical implants, one has to deal with a combination of major challenges in the micro- and nanotechnology field. These are high-throughput patterning, large-area patterning and patterning of, or on, curved surfaces. The first two are closely related. The final one is the most challenging issue in upscaling surface-topographical libraries, particularly in the case of small curvature radii, large arc angles or complex freeform surfaces.

Topographical surface patterns on curved surfaces are considerably distorted when the curvature radii are of similar orders to the dimensions of the surface topographies. Depending on whether the curvature is concave or convex, and whether the surface topographies are pit or pillar types, the regular, predictable distortion results in a corresponding compression or expansion of the pattern grid. A pre-distortion of the pattern via the design file or machine data may be necessary to compensate partly for this.

For beam-based serial patterning, writing time increases with writing area and resolution. Therefore, large areas to be patterned with high resolution result in long writing times. Multiple beams in combination with appropriate writing strategies can reduce the writing time (54, 55). For curved surfaces, beam-based processes need motion systems with rotational axes for tilting the surface to be patterned against the beam. This is necessary to allow the beam to strike the surface and pattern it at a constant angle, or at all. In the case of mask-based parallel patterning, curved surfaces require adaptable or conformal masks to reduce diffraction effects and consequently resolution loss due to small distances between masks and surfaces to be patterned. However, without an appropriate multiaxis motion system, there is still a declination error except where the surface is perpendicular to the incident light rays.

Further issues concerning light- or particle-based lithography on curved surfaces are coating of the surface with resist before its exposure, and pattern transfer after resist development. In most cases, resist coating can no longer be done by spin coating as for flat surfaces. If lower-resolution patterning is acceptable, dip coating can be used. In the case of higher resolution requirements, spray coating has to be employed. After the coating has been applied, isotropic pattern transfer by an appropriate wet etch can access all parts of the curved surface in the same manner. Isotropic etching, however, results in non-straight walls of the etched topographies, under-etching of the resist pattern, and coupling of etching depth and width. Anisotropic pattern transfer by an appropriate plasma etch leads to a declination error where the surface to be patterned is not perpendicular to the incident accelerated ions.

For thermal moulding, the absolute dimensional mismatch between mould and moulded part at the demoulding temperature increases with the side length or diameter of the moulding area, owing to their different coefficients of thermal expansion. Therefore, a large area to be patterned results in large demoulding forces and a high risk of damage. In nanoimprint lithography, however, a large area can be imprinted using a small stamp in multiple steps similar to the 'step and repeat' projection lithography mentioned above.

Because of undercuts, the moulding of parts with curved topographical surfaces requires moulds or moulded parts to be made from highly elastic polymers. Soft lithography enables the moulding of parts with curved topographical surfaces as well as of topographies on existing parts with curved surfaces. The moulding of topographies on curved surfaces needs moulds that are adaptable owing to their elasticity and sheet-like form (56). The moulding of curved topographical surfaces can be done using elastomeric moulds, which in turn are cast from such sheet-like moulds deformed by differential pressure (56, 57).

A way to bypass the generation of topographies on curved surfaces is to create surface topographies on planar thin films by lithography or imprinting, then laminate these patterned films onto curved surfaces, or form them into curved shapes by thermoforming. The topographies on their surfaces are preserved owing to permanent material coherence during forming (58, 59). Thermoforming processes in the medical and pharmaceutical field are mainly used to fabricate packaging products such as trays, clamshells and blisters. Microscale process variants are mainly applied to fabricate thin-walled, flexible biochips.

Implants with a curved topographical surface can also be fabricated by taking a curved part made from a highly elastic material and stretching it flat. Then the part is patterned, for example by transfer from another plane substrate. Finally, the stretched part is released to adopt its original curved shape (60).

8.2.3   Translation to three-dimensional constructs

The fabrication technologies described above have demonstrated their capacity for manufacturing implants with versatile biomaterials and biological patterns such as cell clusters, cell confluent surfaces (which encourage full coverage of cells) and cell patterns, both in 2D and 3D. Nevertheless, like the upscaling of chemical and topographical libraries, upscaling of fabrication technologies faces many challenges. These are mostly related to reproducibility, conservation of material and pattern resolution, time of production and potential toxicity of binding agents.

Moulded implants are easily fabricated in a reproducible manner at low costs and high speed of production, but often need post-processing refinements and do not allow the manufacturing of highly complex shapes. Foaming technologies allow us to produce different medical devices and porous constructs, but offer limited control of the physico-chemical, mechanical and biological properties of the final product. In addition, these technologies create monolithic devices, which cannot be intrinsically modified by additive, subtractive or combinatorial processes owing to their porosity and presence of shadow area. Rapid prototyping technologies are highly reproducible and versatile, while also being able to produce implants with high resolution, but are often slow. The layer-by-layer character of rapid prototyping would facilitate the spatial control of defined surface topographies and chemistries obtained from 2D screening. But convergence and integration of different technological platforms would be required to allow the creation of multiscaled constructs (61). Electrospinning and molecular self-assembly have certainly been shown to be capable of translating 2D hits into 3D systems (62–64), but they are currently limited in terms of reproducibility (e.g. electrospinning is highly dependent on environmental conditions), production time and quantity scales.

Upscaling biofabrication and cell dispensing technologies also faces many challenges. These are mainly related to printing reproducibility and resolution of cell patterns during high-throughput processes (depositing from 10 000 to 1 million droplets per second) and when considering large area patterning (printing cm^2 surfaces). Regarding throughput criteria, the ejection rate and duration of bioprinters must be considered. Since ejection takes place over a few microseconds, one droplet may interfere with the previous one at higher rates. In inkjet printing, increasing ejection rates tend to affect ink refilling into printer heads. In LAB, jet formation, which occurs in 20 μs, may vanish at laser pulse rates above 10 kHz, especially when laser spots are close together. Droplet ejection occurs when a laser-induced cavity collapses. But depositing laser energy close to an existing cavity results in its growth and delays its collapse, thus disturbing the jet

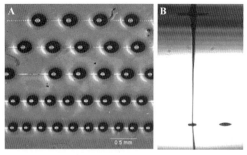

Figure 8.4   Observation of macroscopic bubbles on ribbon, created using LAB. Such unwanted bubbles are obtained when the laser pulse rate is too high relative to the distance between two spots, thus (B) disturbing the jet formation. A, From top to bottom of the picture, the optical scanning velocities are 150, 120, 100, 70 and 50 mm s^{-1}. B, Example of a liquid jet as observed by time-resolved imaging (jetting is from top to bottom). Time delay after laser pulse deposit is 30 μs.

formation (Figure 8.4). In this context, technological solutions for inkjet printing involve setting multiple printing heads in parallel. For LAB, technological improvements might involve the spatio-temporal control of spot distribution on the ribbon using optical scanning systems. High-throughput printing of two contiguous cell-containing dots might be achieved through several laser scanning cycles.

When patterning large-area or large-volume 3D structures, the main issues arise from maintaining a controlled cell density over the process. A cell-level resolution of cell printing (meaning a low volume fraction of bio-ink surrounding the printed cells) at an intermediate throughput (1000–10 000 droplets per second) is already achievable by LAB. Yet, even if the concentration of cells within droplets is roughly proportional to the mean cell concentration in the bio-inks, the number of cells in one droplet remains variable. Such statistical behaviour is amplified by increasing the process duration because of sedimentation in inkjet cartridges and liquid draining onto the ribbon for inkjet and laser assisted bio-printing, respectively. In this context, optical devices allowing droplet selection on-the-fly or controlling the deposition area more accurately could be used for high-throughput purposes. Furthermore, while printing reproducibility is the main prerequisite for upscaling cell-dispensing methods, their capacity to be combined with additional biofabrication technologies, such as two-photon polymerization or microfluidics, might also be crucial to manage the complexity of functional heterogeneous biological constructs.

The fundamental goal of HTS is to isolate conditions with biomedical value. When a promising condition is pinned down, its usefulness is verified through a series of evaluation steps across different size dimensions. In this assembly line leading to scientific and industrial value, parameters that identify positive or negative conditions establish the potential candidates. In an iterative process, false positives or false negatives appear because conditions may not be consistently positive or negative for one parameter across dimensions. A known example of this is that drug candidates show different antitumour potential in 2D versus 3D systems, which can mean that promising results of drugs in 2D are contradicted by results in 3D platforms or vice versa (65). Similarly, HTS

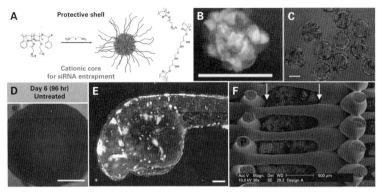

Figure 8.5 HTS in various 3D constructs of increasing size and complexity from the nano- to the macroscale. A, Synthesis of chemically diverse nanoparticles generated a library that was screened for medical potential in hepatocytes (the structure to the left is one example of the molecules going to make up this particle). Adapted with permission from (66). B, Neurospheres cultured on artificial niche arrays and imaged with confocal microscopy express progenitor markers nestin, laminin 1 and Hes5, indicated by different shades of grey. Scale bar: 50 μm. Adapted with permission from (74). C, Three-dimensional microtissues of liver cells stained with cellular Calcein AM viability dye and imaged with phase contrast. Scale bar: 200 μm. Adapted with permission from (68). D, Image of non-treated 7500-cell (A431.H9) spheroid after 96 hours. Scale bar: 200 μm. Adapted with permission from (67). E, Domain-specific expression of enhancer-promoter combinations in transgenic zebra fish embryos. Scale bar: 100 μm. Adapted with permission from (75). F, Bicalcium phosphate particles inserted in the pores of a 3D scaffold of the block copolymer 1000PEOT70PBT30. Scale bar: 500 μm. Adapted with permission from (70).

and the biomedical relevance of its results depend on the dimension where the screening takes place.

HTS is performed at various size scales from the nanoscale (1–100 nm) to the microscale (1–400 μm) (66, 67) depending on the desired read-out (Figure 8.5). Lab-on-a-chip technologies are adapted to the size of the screening unit (e.g. a chemical group on a polymer or cell pellet), thereby optimizing resources and associated experimental costs. Consequently, the screening of chemical groups must be manipulated through nanotechnology (Figure 8.5A), whereas the effect of compounds on cell aggregates (67) must be assessed through microfabrication technology (Figures 8.5B–D). At the microscale, the most common type of platform is the pellet or aggregate of cells which relies on cell-dispensing technologies (Figure 8.5). By using cell aggregates (a 3D rather than 2D approach), new biological discoveries have been made, as cells act differently in 3D (which mimicks the natural environment more closely) than in 2D situations (65, 67, 68). It is important to note that all HTS is currently run at the nano- and microscales whereas the validation of the biomedical potential of hits occurs at the macroscale (>1 mm; Figures 8.5E and F) (69).

The reason HTS takes place at small scales is the cost in time and money. It is prohibitively expensive to perform thousands of experiments on the macroscale. Thus, only 'promising' conditions are tested at the macroscale (69, 70), such as in scaffold platforms, where conditions are carried into the realm of the macroscale to confirm their prospects. A good example of this process can be found in the screening and validation across dimensions (2D versus 3D) of poly(lactic acid), where the screening and validation of chemical, physical and biological aspects of the material began in the early 1970s (71). In 1985, after the first clinical trial (72) and ultimate 3D appraisal, the safety and usefulness of the material was corroborated and authorized for orthopaedic applications.

Following validation, it could be cost-effective to execute HTS on the macroscale, provided well defined and repetitive environments are created and resources are optimized. First, appropriate screening environments in the macroscale can be engineered through, for example, rapid prototyping, in which robotic dispensing of thermoplastics can provide geometrical and size detail and repetition in a suitable scale (>0.5 mm). Furthermore, some of the burden borne by animal experiments could be relocated to manufacturing implantable HTS technology. In doing so, the macroscale evaluation of conditions could be optimized through screening of multiple conditions per animal.

Box 8.1 Classic experiment

To demonstrate a classic approach in terms of upscaling from polymeric libraries to 3D constructs for tissue engineering applications, Khan *et al.* (73) screened the biocompatibility of skeletal cells to 135 polymer blends organized into 960 polymer spots on a glass slide. The screen yielded a blend of poly(lactic acid) and polycaprolactone with affinity for attachment and proliferation of skeletal stem cells. Upon processing, upscaling of the polymer blend to a 3D structure and implantation of the scaffold in mice, the polymer blend was confirmed as a robust template for bone regeneration (Figure 8.6).

To illustrate how surface topographical hits can be currently studied in a low-throughput screening system in 3D, Kumar *et al.* (62) screened the proliferation and differentiation of adult stem cells in a scaffold library composed of two materials, poly (ε-caprolactone) and poly(lactic acid), processed by several techniques which affected the scaffold surface and dimension, thereby decoupling the chemistry from the scaffold architecture. As a result, proliferation and osteogenic markers, gene cluster analysis and cell morphology demonstrated that of the 72 conditions composing the scaffold library, gene expression was significantly different in 3D scaffolds compared with 2D conditions, and that nanofibres alone can stimulate osteogenic differentiation without the presence of osteogenic soluble factors in the culture media (Figure 8.7).

8.3   Future perspectives

High-throughput screening of a chemical library has broad applicability, allowing the identification of materials for specific functions. Studies have focused on cell attachment for particular cell types, controlling adhesion and maintaining the pluripotency of stem cells. This approach may be of interest not only for improving the performance of

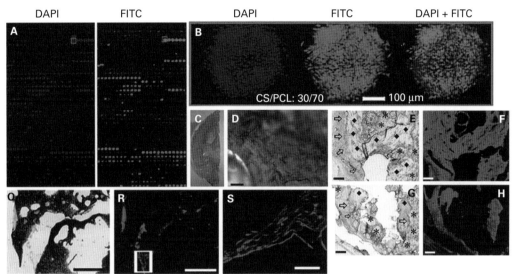

| DAPI | FITC | DAPI | FITC | DAPI + FITC |

Figure 8.6 The adhesion and proliferation potential of human foetal skeletal cells were first screened on a library of synthetic polymers (A, B). The best-performing polymer chemistry, consisting of a PLLA:PCL 80:20 blend, was upscaled into 3D scaffolds cultured in vitro. After 3 weeks (C, D), alkaline phosphatase was expressed, indicating early osteogenic differentiation. After 4 weeks, (Q, R, S) osteocalcin was detected in the seeded scaffolds, among other osteogenic markers, suggesting further osteogenic maturation. After implantation in a segmental femoral defect in immunodeficient mice, (E–H) 3D scaffolds seeded with skeletal stem cells showed newly synthesized bone matrix, (E, G) osteoid formation and (F, H) type I collagen production. Stars indicate remnants of scaffolds; arrows, cells embedded in matrix; and diamonds, osteoid. Scale bars: (Q, R) 500 μM; (D, E, F, G, H, S) 100 μm. Adapted with permission from (73).

medical implants, but ultimately also for identifying materials for sorting populations of cell mixtures from tissues, supporting the growth and controlling the differentiation of cells which are expected to offer enormous therapeutic opportunities. The broad range of biological–material interactions may lead us to discover materials acting as biological factors to regulate cell fate. These innovative solutions may encourage ambitious applications such as the creation of functional tissues and organs from stem cells.

The technique of polymer microarrays has had, and will continue to have, a significant impact on the study and development of new biomaterials with diverse applications. Thousands of new polymers have been tested for their interaction with different cell lines. The approach also appears to offer great potential when allied with cellular biology and the area of tissue engineering.

Although explicit patterning processes making use of focused beams, masks or moulds to create HTS libraries of micro- or nanotopographies are nowadays established methods to study cell–material interactions on 2D substrates, their translation to 3D constructs remains a great challenge. Some steps in this direction have been taken by combining electrospinning with nanoimprinting lithography, but the total surface of the fibre mesh is not yet accessible, owing to fibre curvature and shadow areas. The current technological state of the art allows the generation of micro- and nanotopographies in 3D by processes

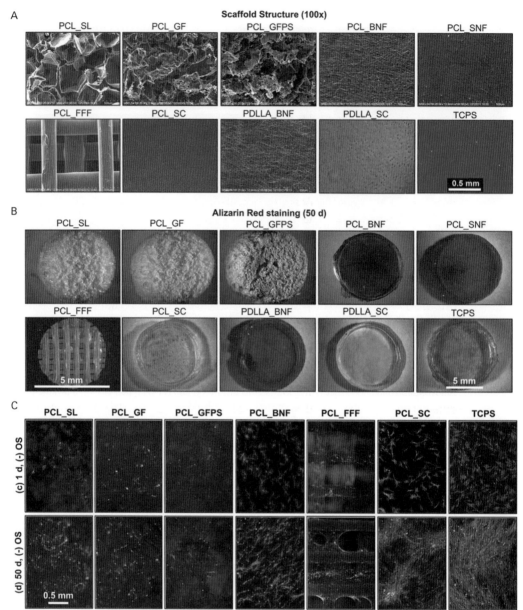

Figure 8.7   Three-dimensional scaffolds were screened to detect the influence of (A) different polymer compositions and architectural properties on osteogenic differentiation. B, Alizarin red staining was observed after 50 days of *in vitro* culture in big nanofibre and small nanofibre scaffolds when cells were cultured in basic media. C, Osteogenic differentiation of mesenchymal stem cells was correlated to cell morphology after 1 and 50 days of cell culture. SL: salt leached; GF: gas foamed; GFPS: gas foamed phase separated; BNF: big nanofibre (~900 nm); SNF: small nanofibre (~300 nm); FFF: freeform fabricated; SC: spin coated; TCPS: tissue culture polystyrene. Adapted with permission from (62).

such as roughening via wet chemical or plasma etching. In these etching processes, amorphous and crystalline regions in metals or polymers etch with a different speed. These processes do not provide the required extent of control over the generated surface topographies, limiting the accuracy, reproducibility, variability or diversity of topographical designs, and ability to generate complex miniaturized topographical libraries. Whereas in explicit patterning processes different topographies mean different designs in a single mask or machining file, in the case of implicit processes (such as wet chemical or plasma etching) multiple process runs are necessary to impart different topographies. However, the implicit processes can be much more powerful than the explicit ones for use on large-area and curved surfaces.

Size does matter in the screening and biomedical evaluation of conditions. Despite the fact that HTS at the microscale is rapidly evolving, particularly with cell pellets, it has yet to gain a foothold at the macroscale. From the biomedical literature, it is increasingly evident that 3D is the natural niche of cells and tissues, so 3D platforms will increasingly improve and boost the pace of preclinical research. It has been partly thanks to robotic dispensing technologies (e.g. cell deposition, polymer deposition) that 3D pellets are relaying HTS into the clinically relevant dimensions. This will continue to be the case because there needs to be repeatability in the amounts deposited and reproducibility of the 3D environment to be able to leap from HTS at the microscale to HTS at the macroscale.

In the past 30 years, nano-, micro- and macroscale platforms have evolved to the point where it is now possible to perform HTS separately in several dimensions, except at the macroscale. However, the integration of screening platforms to allow combined use at different scales, which would certainly accelerate the discovery of medical applications, is yet to come. This integration of screening platforms requires a level of multidisciplinary work on a par with space exploration because of the inherent complexity of the 3D environment of cells and tissues. Consequently, we need to develop tools for the robotic manufacturing of highly defined environments in all dimensions, tools for spatial imaging unlimited by size, and multidisciplinary software for analysis, to integrate nano-, micro- and macroscale HTS into the ultimate platform for biomedical research. Despite this, the current technologies are already leading to the discovery of new avenues to improve the functionality of medical implants and setting us challenges when translating the positive hits found to complex, 3D, macroscopically relevant shapes.

8.4   Snapshot summary

- Chemicals used for combinatorial experiments should represent a wide diversity of chemicals. It is essential to scale up and then compare with the small-scale counterparts to check whether batch-to-batch deviations are affecting the results.
- Screening conditions should echo the application conditions.
- Candidates must be upscaled in several steps from microscale to macroscale to narrow down the best candidates.
- The control of surface topography can span cell and tissue dimensions.

- Robotic dispensing technologies are indispensible for 3D HTS.
- HTS can be upscaled to the microscale where the standard is the generation of cell aggregates.

Further reading

Anderson DG, Levenberg S, Langer R *et al.* Nanoliter-scale synthesis of arrayed biomaterials and application to human embryonic stem cells. Nat Biotechnol. 2004;**22**(7):863–6.

Tourniaire G, Collins J, Campbell S *et al.* Polymer microarrays for cellular adhesion. Chem Commun. 2006;**20**:2118–20.

Zhang R, Liberski A, Sanchez-Martin R *et al.* Microarrays of over 2000 hydrogels – Identification of substrates for cellular trapping and thermally triggered release. *Biomaterials*. 2009;**30** (31):6193–201.

Gehrig J, Reischi M, Kalmar E *et al.* Automated high-throughput mapping of promoter-enhancer interactions in zebrafish embryos. Nat Methods. 2009;**6**(12):911–71.

Gobaa S, Hoehnel S, Roccio M *et al.* Artificial niche microarrays for probing single stem cell fate in high-throughput. Nat Methods. 2011;**8**(11):949–55.

Kumar G, Tison K, Chatterjee K *et al.* The determination of stem cell fate by 3D scaffold structures through the control of cell shape. Biomaterials. 2011;**32**(35):9188–96.

Giselbrecht S, Reinhardt M, Mappes T *et al.* Closer to nature-bio-inspired patterns by transforming latent lithographic images. Adv Mater. 2011;**23**(42):4873–9.

References

1. Scott WL, O'Donnell MJ. Distributed drug discovery, Part 1: Linking academia and combinatorial chemistry to find drug leads for developing world diseases. J Combin Chem. 2009;**11** (1):3–13.
2. Anderson DG, Levenberg S, Langer R. Nanoliter-scale synthesis of arrayed biomaterials and application to human embryonic stem cells. Nature Biotechnol. 2004; **22**(7):863–6.
3. Thaburet JFO, Mizomoto H, Bradley M. High-throughput evaluation of the wettability of polymer libraries. Macromol Rapid Commun. 2004; **25**(1):366–70.
4. Zhang R, Liberski A, Sanchez-Martin R, Bradley M. Microarrays of over 2000 hydrogels – Identification of substrates for cellular trapping and thermally triggered release. Biomaterials. 2009;**30**(31):6193–201.
5. Madou MJ. *Fundamentals of Microfabrication and Nanotechnology* 3rd edn: CRC Press; 2011.
6. Menz W, Mohr J, Paul O. *Microsystem Technology* Revised edn. Cambridge: Wiley-VCH; 2001.
7. Xia YN, Whitesides GM. Soft lithography. Angew Chem Int Ed. 1998; **37**(5):551–75.
8. Zaidi SH, Brueck SRJ. Multiple-exposure interferometric lithography. J Vac Sci Technol B. 1993;**11**(3):658–66.
9. Giboz J, Copponnex T, Mele P. Microinjection moulding of thermoplastic polymers: a review. J Micromech Microeng. 2007;**17**(6):R96–R109.

10. Worgull M. *Hot Embossing – Theory and Technology of Microreplication*. William Andrew; 2009.

11. Carvalho BL, Schilling EA, Schmid N, Kellogg GJ, eds. Soft embossing of microfluidic devices. Proc 7th International Conference on Miniaturized Chemical and Biochemical Analysis Systems; 2003; Squaw Valley.

12. Russo AP, Apoga D, Dowell N *et al.* Microfabricated plastic devices from silicon using soft intermediates. Biomed Microdevices. 2002;**4**(4):277–83.

13. Chou SY, Krauss PR, Renstrom PJ. Imprint lithography with 25-nanometer resolution. Science. 1996;**272**(5258):85–7.

14. Murali M, Yeo SH. Rapid biocompatible micro device fabrication by micro electro-discharge machining. Biomed Microdevices. 2004;**6**(1):41–5. Epub 2004/08/17.

15. Schuster R, Kirchner VV, Allongue P, Ertl G. Electrochemical micromachining. Science. 2000;**289**(5476):98–101. Epub 2000/07/07.

16. Coombs AG, Lawrence RB, Davies RM. Rotational moulding in the production of prostheses. Prosthet Orthot Int. 1985;**9**(1):31–6. Epub 1985/04/01.

17. Fonseca A, Inacio N, Kanagaraj S, Oliveira MS, Simoes JA. The use of Taguchi technique to optimize the compression moulding cycle to process acetabular cup components. J Nanosci Nanotechnol. 2011;**11**(6):5334–9. Epub 2011/07/21.

18. Freischlag JA, Moore WS. Clinical experience with a collagen-impregnated knitted Dacron vascular graft. Ann Vasc Surg. 1990;**4**(5):449–54. Epub 1990/09/01.

19. Sahoo S, Ouyang H, Goh JC, Tay TE, Toh SL. Characterization of a novel polymeric scaffold for potential application in tendon/ligament tissue engineering. Tissue Eng. 2006;**12**(1):91–9.

20. Hou Q, Grijpma DW, Feijen J. Porous polymeric structures for tissue engineering prepared by a coagulation, compression moulding and salt leaching technique. Biomaterials. 2003;**24**(11):1937–47. Epub 2003/03/05.

21. Sproule TL, Lee JA, Li HB, Lannutti JJ, Tomasko DL. Bioactive polymer surfaces via supercritical fluids. J of Supercritical Fluids. 2004;**28**(2–3):241–8.

22. Hollister SJ. Porous scaffold design for tissue engineering. Nat Mater. 2005;**4**(7):518–24.

23. Pfister A, Landers R, Laib A *et al.* Biofunctional rapid prototyping for tissue-engineering applications: 3D bioplotting versus 3D printing. J Polym Sci Part A: Polym Chem. 2004;**42**:624–38.

24. Engelhardt S, Hoch E, Borchers K *et al.* Fabrication of 2D protein microstructures and 3D polymer–protein hybrid microstructures by two-photon polymerization. Biofabrication. 2011;**3**(2):025003. Epub 2011/05/13.

25. Li D, Xia YN. Electrospinning of nanofibers: Reinventing the wheel? Adv Mater. 2004;**16**(14):1151–70.

26. Zhang SG. Fabrication of novel biomaterials through molecular self-assembly. Nat Biotechnology. 2003;**21**(10):1171–8.

27. Klebe RJ. Cytoscribing: a method for micropositioning cells and the construction of two- and three-dimensional synthetic tissues. Exp Cell Res. 1988;**179**(2):362–73. Epub 1988/12/01.

28. Boland T, Xu T, Damon B, Cui X. Application of inkjet printing to tissue engineering. Biotechnol J. 2006;**1**(9):910–7. Epub 2006/08/31.

29. Nakamura M, Iwanaga S, Henmi C, Arai K, Nishiyama Y. Biomatrices and biomaterials for future developments of bioprinting and biofabrication. Biofabrication. **2**(1):014110. Epub 2010/09/03.

30. Pernagallo S, Diaz-Mochon JJ, Bradley M. A cooperative polymer-DNA microarray approach to biomaterial investigation. Lab Chip. 2009;**9**(3):397–403.

31. Saunders R, Gough J, Derby B. Delivery of human fibroblast cells by piezoelectric drop-on-demand inkjet printing. Biomaterials. 2008;**29**(2):193–203.

32. Nair K, Gandhi M, Khalil S *et al.* Characterization of cell viability during bioprinting processes. Biotechnol J. 2009;**4**(8):1168–77. Epub 2009/06/10.

33. Cui XF, Boland T. Human microvasculature fabrication using thermal inkjet printing technology. Biomaterials. 2009;**30**(31):6221–7.

34. Xu T, Jin J, Gregory C, Hickman JJ, Boland T. Inkjet printing of viable mammalian cells. Biomaterials. 2005;**26**(1):93–9.

35. Fedorovich NE, De Wijn JR, Verbout AJ, Alblas J, Dhert WJ. Three-dimensional fiber deposition of cell-laden, viable, patterned constructs for bone tissue printing. Tissue Eng Part A. 2008;**14**(1):127–33. Epub 2008/03/13.

36. Mironov V, Visconti RP, Kasyanov V *et al.* Organ printing: Tissue spheroids as building blocks. Biomaterials. 2009;**30**(12):2164–74.

37. Guillemot F, Souquet A, Catros S, Guillotin B. Laser-assisted cell printing: principle, physical parameters versus cell fate and perspectives in tissue engineering. Nanomedicine. 2010;**5** (5):507–15.

38. Bohandy J, Kim BF, Adrian FJ. Metal-deposition from a supported metal-film using an excimer laser. J Appl Phys. 1986;**60**(4):1538–9.

39. Brey DM, Motlekar NA, Diamond SL *et al.* High-throughput screening of a small molecule library for promoters and inhibitors of mesenchymal stem cell osteogenic differentiation. Biotechnol Bioeng. 2011;**108**(1):163–74.

40. Martis EA, Radhakrishnan R, Badve RR. High-throughput screening: the hits and leads of drug discovery – an overview. J Appl Pharmaceut Sci. 2011;**1**:2–10.

41. Yamada Y, Kobayashi T. Utilization of combinatorial method and high-throughput experimentation for development of heterogeneous catalysts. J Jap Petrol Inst. 2006;**49**(4):157–67.

42. Zhang K, Liu QF, Liu QA, Shi Y, Pan YB. Combinatorial Optimization of (Y(x)Lu(1–x–y))(3) Al(5)O(12):Ce(3y) green-yellow phosphors. J Combin Chem. 2010;**12**(4):453–7.

43. Derda R, Musah S, Orner BP *et al.* High-throughput discovery of synthetic surfaces that support proliferation of pluripotent cells. J Am Chem Soc. 2010;**132**(4):1289–95.

44. Jung G. *Combinatorial Chemistry: Synthesis, Analysis, Screening.* Wiley-VCH; 1999.

45. Potyrailo RA, Ezbiansky K, Chisholm BJ *et al.* Development of combinatorial chemistry methods for coatings: High-throughput weathering evaluation and scale-up of combinatorial leads. J Combin Chem. 2005;**7**(2):190–6.

46. Potyrailo R, Rajan K, Stoewe K *et al.* Combinatorial and high-throughput screening of materials libraries: review of state of the art. ACS Comb Sci. 2011;**13**(6):579–633. Epub 2011/06/08.

47. Diaz-Mochon JJ, Tourniaire G, Bradley M. Microarray platforms for enzymatic and cell-based assays. Chem Soc Rev. 2007;**36**(3):449–57.

48. Hook AL, Anderson DG, Langer R *et al.* High-throughput methods applied in biomaterial development and discovery. Biomaterials. 2010;**31**(2):187–98.

49. Hansen A, McMillan L, Morrison A, Petrik J, Bradley M. Polymers for the rapid and effective activation and aggregation of platelets. Biomaterials. 2011;**32**(29):7034–41.

50. Hay DC, Pernagallo S, Diaz-Mochon JJ *et al.* Unbiased screening of polymer libraries to define novel substrates for functional hepatocytes with inducible drug metabolism. Stem Cell Res. 2011;**6**(2):92–102.

51. Tare RS, Khan F, Tourniaire G, Morgan SM, Bradley M, Oreffo ROC. A microarray approach to the identification of polyurethanes for the isolation of human skeletal progenitor cells and augmentation of skeletal cell growth. Biomaterials. 2009;**30**(6):1045–55.

52. Tourniaire G, Collins J, Campbell S *et al.* Polymer microarrays for cellular adhesion. Chem Commun. 2006;**20**:2118–20.

53. Zhang R, Liberski A, Khan F, Diaz-Mochon JJ, Bradley M. Inkjet fabrication of hydrogel microarrays using in situ nanolitre-scale polymerisation. Chem Commun. 2008;**11**:1317–19.

54. Chang THP, Mankos M, Lee KY, Muray LP. Multiple electron-beam lithography. Microelectr Eng. 2001;**57**–8:117–35.

55. Slot E, Wieland MJ, Boer Gd *et al.*, eds. MAPPER: high-throughput maskless lithography. Proc SPIE Conference on Emerging Lithographic Technologies XII; 2008; San Jose, CA.

56. Xia YN, Kim E, Zhao XM *et al.* Complex optical surfaces formed by replica moulding against elastomeric masters. Science. 1996;**273**(5273):347–9.

57. Hoffman JM, Shao J, Hsu CH, Folch A. Elastomeric moulds with tunable microtopography. Adv Mater. 2004;**16**(23–24):2201.

58. Giselbrecht S, Reinhardt M, Mappes T *et al.* Closer to nature: bio-inspired patterns by transforming latent lithographic images. Adv Mater. 2011;**23**(42):4873–9. Epub 2011/09/22.

59. Truckenmuller R, Giselbrecht S, Escalante-Marun M *et al.* Fabrication of cell container arrays with overlaid surface topographies. Biomed Microdevices. Epub 2011/11/04.

60. Ko HC, Stoykovich MP, Song JZ *et al.* A hemispherical electronic eye camera based on compressible silicon optoelectronics. Nature. 2008;**454**(7205):748–53.

61. Moroni L, de Wijn JR, van Blitterswijk CA. Integrating novel technologies to fabricate smart scaffolds. J Biomater Sci. 2008;**19**(5):543–72.

62. Kumar G, Tison CK, Chatterjee K *et al.* The determination of stem cell fate by 3D scaffold structures through the control of cell shape. Biomaterials. 2011;**32**(35):9188–96. Epub 2011/09/06.

63. Dankers PY, Harmsen MC, Brouwer LA, van Luyn MJ, Meijer EW. A modular and supramolecular approach to bioactive scaffolds for tissue engineering. Nat Mater. 2005;**4**(7):568–74. Epub 2005/06/21.

64. Capito RM, Azevedo HS, Velichko YS, Mata A, Stupp SI. Self-assembly of large and small molecules into hierarchically ordered sacs and membranes. Science. 2008;**319**(5871):1812–6. Epub 2008/03/29.

65. Kunz-Schughart LA, Freyer JP, Hofstaedter F, Ebner R. The use of 3-D cultures for high-throughput screening: The multicellular spheroid model. J Biomol Screening. 2004;**9**(4):273–85.

66. Siegwart DJ, Whitehead KA, Nuhn L *et al.* Combinatorial synthesis of chemically diverse core–shell nanoparticles for intracellular delivery. Proc Natl Acad Sci USA. 2011;**108**(32):12996–3001.

67. Tung YC, Hsiao AY, Allen SG *et al.* High-throughput 3D spheroid culture and drug testing using a 384 hanging drop array. Analyst. 2011;**136**(3):473–8.

68. Chen AA, Underhill GH, Bhatia SN. Multiplexed, high-throughput analysis of 3D microtissue suspensions. Integr Biol. 2010; **2**(10):517–27.

69. Liu XH, Jin XB, Ma PX. Nanofibrous hollow microspheres self-assembled from star-shaped polymers as injectable cell carriers for knee repair. Nat Mater. 2011;**10**(5):398–406.

70. Moroni L, Hamann D, Paoluzzi L *et al.* Regenerating articular tissue by converging technologies. Plos One. 2008;**3**(8).

71. Kulkarni RK, Moore EG, Hegyeli AF, Leonard F. Biodegradable poly(lactic acid) polymers. J Biomed Mater Res. 1971;**5**(3):169–81. Epub 1971/05/01.

72. King JB, Bulstrode C. Polylactate-coated carbon fiber in extra-articular reconstruction of the unstable knee. Clin Orthop Relat Res. 1985(196):139–42. Epub 1985/06/01.

73. Khan F, Tare RS, Kanczler JM, Oreffo RO, Bradley M. Strategies for cell manipulation and skeletal tissue engineering using high-throughput polymer blend formulation and microarray techniques. Biomaterials. 2010;**31**(8):2216–28. Epub 2010/01/09.

74. Gobaa S, Hoehnel S, Roccio M *et al* Artificial niche microarrays for probing single stem cell fate in high-throughput. Nat Methods.**8**(11):949–55. Epub 2011/10/11.

75. Gehrig J, Reischl M, Kalmar E *et al*. Automated high-throughput mapping of promoter-enhancer interactions in zebrafish embryos. Nat Methods. 2009;**6**(12): 911–71.

9 Development of materials for regenerative medicine: from clinical need to clinical application

Charlène Danoux, Rahul Tare, James Smith, Mark Bradley, John A. Hunt, Richard O. C. Oreffo and Pamela Habibovic

Scope

Given the demographic challenges of an ageing population combined with rising patient expectation and the growing emphasis placed on cost containment by healthcare providers, economic regenerative medicine approaches for regeneration of damaged and diseased organs and tissues are a major clinical and socio-economic need. The scope of this chapter is to use skeletal regeneration as the exemplar to discuss classical and high-throughput screening approaches to biomaterials development for regenerative medicine, including choice and design of materials based on clinical need, biological assessment and regulatory issues.

9.1 Basic principles: development of materials for regenerative medicine

The increase in an ageing population in developed countries is accompanied by a growing need for replacement and repair of damaged organs and tissues. Transplantation of the patient's own tissue is still considered the gold standard in many applications, but limited availability, and complications associated with harvesting of the so-called autograft, are becoming an important drawback. Tissues and organs from human or animal donors present issues of disease transmission and functional failure. Alternative strategies, based on biological growth factors, cell therapy and tissue-engineered constructs, are being explored as alternatives to the patient's own tissue, but their use is hampered by biological instability and high costs. These issues demonstrate the need for strategies based on biomaterials, which are often synthetic, and thus less prone to instability problems. In addition, the fact that (synthetic) biomaterials can often be produced in large quantities and thus be available off-the-shelf is an important advantage when coping with an increasing need for regenerative approaches.

Following a consensus by experts in the field, a biomaterial is defined as a 'nonviable material used in a medical device to interact with biological system' (1). A number of

Materiomics: High-Throughput Screening of Biomaterial Properties, ed. Jan de Boer and Clemens van Blitterswijk. Published by Cambridge University Press. © Cambridge University Press 2013.

other, non-medical applications also include an interaction between a material and a biological system, for example materials used to produce diagnostic gene arrays or equipment to process biomolecules in various biotechnological applications, and can therefore be considered biomaterials. The focus of this chapter, however, is to describe the process from synthesis of a material to clinical application and to discuss how high-throughput screening of materials can make such a process more efficient.

9.1.1 The changing role of materials in medicine

Historically, the primary application of materials in medicine was in implants, the role of which was to replace and take over the function of a diseased/damaged organ or tissue. From this perspective, the choice of materials for and design of implants are such that the natural organ or tissue that needs replacement is mimicked as closely as possible. The implant needs to be accepted by the *in vivo* surrounding and perform its function without actively affecting the functioning of other organs or tissues. Heart valve prostheses, for example, which can be made from various materials including stainless steel and carbon, instantly recover cardiac function upon implantation. The intraocular lenses used to replace a natural lens when it becomes cloudy through cataract formation are made of poly(methyl methacrylate), silicone rubber or a hydrogel, and their insertion results in immediate restoration of vision (see also Classic experiments). Dental implants, usually made of titanium, provide an artificial tooth anchor upon which a crown is placed. Hip joint prosthesis is a good example of an implant where different materials need to be combined in order to restore the function of the natural joint: a metallic stem is required in order for the implant to withstand high levels of mechanical stress, a polymeric liner allows for smooth movement of the femoral head inside the acetabular cup, and PMMA cement is needed to fix the stem inside the femur upon placement (2). This serves to demonstrate that the complexity of the organ or tissue that needs to be replaced determines the complexity of the implant from the perspective of both raw material and implant design.

Box 9.1 Classic experiment

Development of intraocular lens

Sir Harold Ridley MD (1906–2001) was a pioneer of intraocular lens (IOL) development. IOL are meant to replace the crystalline lens of the eye in patients suffering from cataract, which is a clouding of the lens leading to partial or complete blindness. Surgery procedures for this condition were developed in ancient times, but the sole removal of the crystalline lens resulted in a blurry and unfocused vision. In the 1930s, Harold Ridley, an ophthalmic surgeon, recognized the need for a synthetic lens. After World War II, Ridley treated injured pilots from the Royal Air Force, and observed the interesting consequences of fragments of Hurricane cockpit canopies lodged in the eyes of the pilots. He noted that no reaction had been triggered by the fragments, and that the material seemed to be well tolerated by the human eye. Ridley's observations would today relate to the concept of biocompatibility. He researched the canopy material, which was poly(methyl methacrylate) (ICI Perspex, also known under the trademark name of Plexiglas), and used this material to design the first IOL.

In 1949, Ridley performed his first implantation of IOL in a cataract patient at Saint Thomas' Hospital in London. Although the outcome of the surgery was very encouraging, he faced strong opposition from the medical community, which fiercely disapproved of implantation of a foreign material in the eye. Between 1949 and 1966, approximately a thousand IOL were implanted by Ridley in patients in Europe and in the United States, in order to prove the robustness of the technique. The success rate of these surgeries was evaluated to be around 70%. However, it was not until the 1980s that this technique was accepted as a general treatment for cataract. Nowadays, tens of millions of IOL are implanted each year, with successful restoration of sight in cataract patients. Harold Ridley's talent and merit were first acknowledged in 1986, with his election to the Fellowship of the Royal Society, and in 2000, he received the honour of being knighted by Queen Elizabeth II (3, 4).

Current strategies in treatment of damaged organs and tissues require a more active role from implanted materials. Instead of, relatively passively, replacing and thus taking over the role of an organ or tissue, the goal is now to repair the natural tissue therewith restoring its function. This regenerative medicine approach poses different requirements for the biomaterials to be implanted. They must actively facilitate new tissue formation and, in some cases, even instruct the *in vivo* environment to exert a certain function. Such an active role is often associated with changes in the material itself after implantation, which needs to be predicted before the materials can be used in the clinic. The changing role of materials in regenerative medicine is illustrated in Figure 9.1, in which two examples of prosthetic heart valves are shown.

9.1.2   Classic approach to materials development

Independent of whether a material will be used in a 'passive' device, or as a tool to regenerate an organ or tissue, a number of choices need to be made before and during development. These choices are in the first place determined by the role the material needs to play in the body. In addition, the process of implantation itself, which needs to be efficient, safe and minimally invasive for the patient, will also determine certain properties of the materials. In bone regeneration, for example, the biomaterial should be able to initiate and/or facilitate new bone formation and (temporarily) provide mechanical support to bone defects (7). Constructs for treatment of peripheral nerve injuries should have an oriented morphology, in order to guide regeneration of a damaged nerve (8). Materials used in peripheral stent grafts should provide a mechanical barrier to prevent intravascular pressure from being transmitted to the weakened wall of the aneurysm (9).

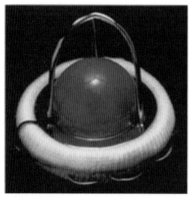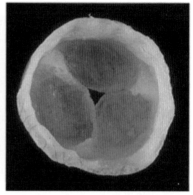

Figure 9.1   Examples of prosthetic heart valves. A, Caged-ball (Starr–Edwards) valve (1960). This artificial valve proved to have excellent durability; however, it also triggered very high thrombogenicity (5). B, Trileaflet valve scaffold, made of a poly(glycolic acid)/poly(4-hydroxybutyrate) composite, seeded with myofibroblasts and endothelial cells. The construct was implanted in an ovine model for 20 weeks and showed promising results: the scaffold is able to provide support and guidance for tissue growth and to resorb to create space for a regenerated valve (6).

From this it is obvious that development of a biomaterial is a highly multidisciplinary process, where knowledge of materials science, engineering, biology and anatomy needs to be combined, while taking into account clinical requirements.

By taking development of bone graft substitutes as an example, we will now discuss the classical process of biomaterials development, based on rational design, including the material choice and design, pre-clinical *in vitro* and *in vivo* tests of safety and functionality, and regulations that are important in order to bring a material to clinical application.

Material choice and implant design

As mentioned before, what determines the properties of a material intended for implantation is the characteristics of the tissue/organ that needs to be regenerated and the function the material needs to fulfil. The role of synthetic bone graft substitutes is to restore a bone defect, caused for example by trauma, tumour removal, extensive resorption or congenital diseases.

Bone is a highly specialized form of connective tissue that is nature's provision for an internal support system in higher vertebrates. Bone provides for the attachment of muscles and tendons essential for locomotion, protects the vital organs of the cranial and thoracic cavities, and encloses the blood-forming elements of the bone marrow. In addition to these mechanical functions, bone plays an important metabolic role as a store of calcium and phosphate, which can be drawn upon when needed in the homeostatic regulation of calcium and phosphate in blood and other fluids of the body (10). By weight, bone contains approximately 60% mineral (mainly type AB carbonated calcium phosphate apatite), 10% water and about 30% organic matrix (predominantly type I collagen, as well as proteoglycans and non-collageneous proteins, such as osteocalcin, osteopontin, osteonectin, bone sialoprotein, decorin and biglycan) (11, 12). Morphologically, there are two forms of bone: cortical (compact) bone and cancellous (trabecular) bone. Compact bone, which is rigid and dense, is found mainly in the middle shaft of long bones or shells of other bones. Found predominantly in epiphysis, ribs and spine, cancellous bone has a highly porous structure (>75%), with numerous small bone trusses or trabeculae interconnected with each other, and tends to orient along the principal directions in adaptation to the external loading environment. In the formation of the skeleton, mesenchymal cells aggregate and form condensates of loose mesenchymal tissue, prefiguring the skeletal elements. Within these aggregates, cells may differentiate into osteoblasts, when in association with adequate vascularization, thereby directly initiating ossification, which eventually results in either compact or cancellous bone (intramembranous bone formation). Alternatively, condensates of mesenchymal cells can differentiate into chondrocytes in an avascular environment, producing cartilage which is eventually replaced by bone (endochondral bone formation) (13).

The role of a bone graft substitute is to aid repair of a bone defect or to restore bone volume. Therefore, a bone graft substitute should be osteoconductive, meaning that it should allow migration of potentially osteogenic cells to the site of (orthotopic) implantation (14), which eventually results in deposition of bone on the surface of a bone graft substitute. Ideally, a bone graft substitute should also be osteoinductive, a process defined as the induction of undifferentiated inducible osteoprogenitor cells that are not yet committed to the osteogenic lineage to form osteoprogenitor cells (15), meaning that the material should induce bone formation. In order to restore the function of defected bone as efficiently as possible, the process of bone formation and the resulting structural properties of the newly formed bone should be equal or at least similar to what is occurring in the body. For example, when a bone graft substitute is used to treat a calvarial defect, bone formation should follow the direct pathway, whereas for the repair of a long bone, for example femoral defect, the pathway of bone formation should be endochondral. Similarly, bone formed in a femoral defect should be cortical, in contrast to the bone found in the interior of vertebrae, which is predominantly cancellous, which will also determine its mechanical properties.

In classic, candidate approaches towards biomaterials development, a logical starting point is to mimic the composition of natural tissue. As mentioned before, the mineral component of bone is a calcium phosphate ceramic, and it is therefore not surprising that the most widely applied bone graft substitutes are based on calcium phosphates (7). The exact composition (for example calcium to phosphorus ratio, presence of other additives such as fluorine, magnesium) varies, which determines properties of these materials such as degradation rate and mechanical strength, but the rationale in all cases is to mimic bone mineral or its precursors. That such a biomimetic approach is logical is witnessed by the fact that the majority of calcium phosphate ceramics are biocompatible and bioactive in terms of osteoconductivity, and in some cases even osteoinductivity (16).

Besides chemical composition, other physical properties, such as macro- and micro-structure, are often mimicked too. In most calcium phosphate ceramics, one aims at introducing an interconnected porous structure into the material, which allows infiltration by cells, blood vessels, and therewith nutrients and oxygen as well as in growth of *de novo* tissue. A great number of studies are therefore focused on determining optimal pore size, importance of interconnectivity, and methods to obtain these (17). Finally, surface properties are often used as a tool to control interactions of the material with its *in vivo* environment. For example, roughness of the surface, its size, and the presence of micro- and nanofeatures are all of importance in determining initial contact interaction with body fluids, including immune response, and protein adsorption, as well as subsequent cell proliferation, differentiation and matrix formation (18, 19, 20).

Apart from these material properties, which are designed in such a way that upon implantation they lead to bone formation that is similar to the natural process, there are also needs or wishes from the clinicians that should be satisfied. For example, in many cases, an orthopaedic or maxillo-facial surgeon would prefer a bone graft substitute that is mouldable or injectable. This is why calcium phosphate cements have been developed (21), and why there is much research performed on mixing granules of calcium phosphate ceramics with polymeric gels, etc., to allow injectability (22, 23). The restrictions that are associated with such modifications are that it is much more difficult to introduce pores into cement, and besides, the size of ceramic granules needs to be modified to allow injectability. Thus, a compromise is to be reached. Furthermore, while ceramic materials based on calcium phosphates perform very well *in vivo*, they are intrinsically brittle, which means that they cannot be used in loaded applications. This of course is an important disadvantage, considering the load-bearing functions of bone. In order to improve mechanical properties, other materials, for example polymers or ceramic/polymer mixtures, are considered, but this usually is a guarantee for a decrease in bioactivity.

All this is to demonstrate how difficult it is to reproduce the complexity of natural tissue and that often many choices and compromises need to be made along the way of material development. This also shows why, despite over 60 years of intensive work on synthetic bone graft substitutes, there is still no truly comprehensive alternative to patient's own bone.

Biological assessment of biomaterials

Once a material is designed and developed, a lengthy process of tests needs to be followed in order to bring the potential bone graft substitute to the clinic. In general, this process starts with pre-clinical *in vitro* assays using a cell- or an organ culture system. It should be emphasized that classical *in vitro* assays have initially been developed to study the influence of growth factors and hormones on attachment, proliferation, differentiation and mineralization of cells, for example. These assays have been implemented in the field of biomaterials research without significant modifications. What is, however, often ignored is that the *in vitro* setting may significantly be changed by the presence of a material, for example because of interaction between the material and the cell culture medium, and that this change is highly dependent on the type of material tested. As a consequence, it is difficult to distinguish the true responses of cells to a material from the unwanted response of cells to the changes in the cell culture environment (24). Needless to say, *in vitro* assays are attractive because of their simplicity, but, at the same time, this simplicity is an important limitation, in particular because presence of a material adds to the complexity of the cell culture system.

In vitro cell- and organ culture assays are in the first place used to investigate the 'safety' of the material in terms of cytotoxicity and biocompatibility. Three main cell culture assays are used for evaluating biocompatibility: direct contact, agar diffusion and elution.The L-929 mouse fibroblast cell line has been most extensively used for testing biomaterials. In addition to the tests that focus on safety aspects of biomaterials and are independent of the intended application of the material, *in vitro* bone formation assays are used to predict the performance of the material *in vivo* in its role of a bone graft substitute, meaning that one aims at getting an idea about osteoconductive and osteoinductive potential of the material. For this, again, a rational choice of cell type and cell culture conditions is made in an attempt to mimic the *in vivo* environment. In the case of bone graft substitutes, this often means that one uses osteoblasts (either as cell lines such as MC3T3-E1 or MG63, or primary osteoblasts) and mesenchymal stromal cells, known to possess the ability to differentiate into the osteogenic lineage and form bone, as a model of choice. With components of cell culture medium, specific stimuli acting on osteogenic differentiation or mineral production are added in order to simulate various phases of bone formation. Read-out of such assays is usually cell proliferation, expression of markers of osteogenesis at gene and protein level, production of extracellular matrix and mineral formation. Besides these assays which are directed to study potential bone forming capacity, other assays relevant to bone formation and remodelling, such as angiogenesis (new blood vessel formation) and osteoclastogenesis (degradation by bone-resorbing cells), are performed too, with relevant cell types (endothelial cells and osteoclasts, respectively) (24). Similar to orthopaedic research, functional *in vitro* assays with either primary cells or cell lines are used to test the applicability of materials in other applications. For example, the ability of primary chondrocytes to retain their chondro-genic phenotype and the ability to produce type II collagen in culture is an important assay in cartilage research (25). In cardiac research, the ability of cells to differentiate into functional and beating cardiomyocytes is of great relevance for treatment of infarcted myocardium (26). An example of a relevant assay in nervous system regeneration research is the ability of Schwann cells to provide a growth substrate to central nervous system axons (8).

Although *in vitro* assays present a relatively easy way to test biological performance of potential bone graft substitutes, the truth is that they are too simple and therefore often fail to predict the *in vivo* behaviour of the materials. This is why the next step in testing these materials is pre-clinical *in vivo* studies involving animals. As reviewed by Jansen (27), the first test following *in vitro* assays is *in vivo* compatibility of materials for short and prolonged periods of time. Soft tissue implantation is an attractive model to study the safety of the materials in terms of toxicity and carcinogenicity, as it is rather inexpensive, readily available and yet relevant, as many materials used as bone graft substitutes come in contact with subcutaneous tissue, muscles, fasciae and tendons. The two most frequently used soft tissue models are subcutaneous and intramuscular implantation. It has been shown that the biocompatibility response of implant materials can differ between the two test sites, owing to differences in vascularization, regenerative capacity and intrinsic stress. The selection of a suitable animal for biocompatibility testing is another complex issue. Mice, rats and rabbits are most often used for soft tissue implantations. The advantage of these relatively small animals is their availability and low cost. However, their metabolic and wound healing properties differ from those of large animals and humans. In addition to testing safety of biomaterials, soft tissue models are needed to study osteoinductive potential materials, as a true proof of osteoinductivity is bone formation at heterotopic sites. An and Friedman gave an overview of the frequently used soft tissue models (e.g. subcutaneous, intramuscular, intraperitoneal and mesentery) to assay osteogenicity prior to orthotopic implantation (28).

Finally, functional *in vivo* tests are performed to test the capacity of a bone graft substitute in its true role, and thus the ability to close a bone defect, help regenerating bone and restoring its function. Four types of defect are typically used: calvarial-, long bone (or mandible) segmental-, partial cortical- (e.g. cortical window, wedge defect or transcortical drill hole) and cancellous bone defects. Various animals are used for these functional assays, including rats, sheep, dogs, goats and non-human primates, and ideally, bone metabolism and mechanical loading should be highly representative for those found in humans. An elegant overview of different animal models of bone defect repair is provided by An and Friedman (29). While bone metabolism and mechanical loading of the skeleton determine which animals are most suitable for testing materials in orthopaedic applications, in biomaterials intended for regeneration of other organs and tissues other characteristics of an animal are more relevant. In cardiac research, for example, cardiovascular physiology, ease of vascular access and low body fat have made canines the model of choice for cardiac and peripheral vascular studies for many years, although more recently swine and sheep have been used too (30). To study repair of articular cartilage defects, the thickness of articular cartilage in, for example, femoral condyle, the thickness of the defect as well as joint size and loading conditions determine the suitability of an animal model. Regarding the thickness of cartilage and the volume and nature of the defect, the horse shows the best resemblance with humans. Regarding loading, goats and sheep seem a more suitable model (31).

From R&D results to clinical application

Biomaterials are subjected to a regulatory process, similar to that of therapeutic drugs, before they are allowed to be marketed. This process is conducted by the manufacturer, who has to provide sufficient evidence of the biomaterial safety and efficacy in the intended application, to the regulatory authority. For the US market, the Food and Drug Administration (FDA) is responsible for the evaluation of the applications. To help the manufacturers building their report – also called 510(k) submission – the FDA provides guidelines on tests that need to be performed to provide the required data (e.g. FDA documents for bone void and dental bone grafting devices). There are no specific or quantified requirements to meet. However, it is advised to establish a detailed comparison between the new biomaterial and a material already marketed or approved.

Depending on what is already known about the biomaterial entering the regulatory process, the manufacturer has to choose between a traditional, an abbreviated or a specific 510(k) submission (see The New 510(k) Paradigm, Figure 9.2). For a traditional submission, requirements are described in 21 CFR 807, subpart E. An abbreviated submission has to include a description of the device and its intended use, a description of the device design

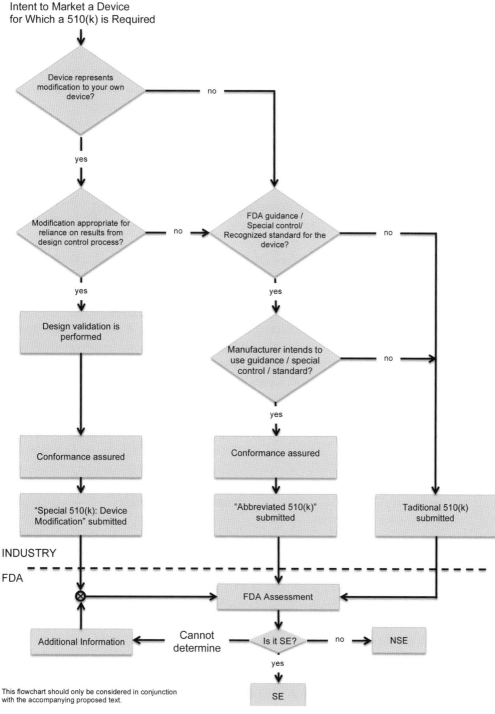

Figure 9.2 The new 510(k) paradigm showing the regulatory process for obtaining FDA approval for a medical device.

and performances, an identification of the risk analysis method, and a discussion of the device characteristics. For specific submissions, precise guidelines are provided; 'Class II Special Controls Guidance document: Resorbable Calcium Salt Bone Void Filler Device', for example, gives a list of physical properties (porosity, size, shape, surface area, mass to volume ratio) and chemical analysis (X-ray diffraction and Fourier transform infrared spectroscopy) to be provided. It also integrates animal testing to evaluate performances such as bone formation, biomaterial resorption and mechanical properties. To facilitate the evaluation of 510(k) submissions, FDA advises the use of standards, such as ISO-10993 'Biological Evaluation of Medical Devices', or American Society for Testing and Materials (ASTM) standards, which give an insight on how to perform the different tests.

9.2   Relation to materiomics: from high-throughput screening to large animal studies and beyond

The lengthy and expensive process from material design to clinical trial involves a number of assumptions and choices along the way, some of which may be incorrect, meaning that there is a fair chance that potentially excellent bone graft substitutes will be missed. In the absence of iterative and high-throughput robust analysis, the current portfolio of bone graft substitutes has remained remarkably unchanged for years. Current research strategies therefore aim at replacing the traditional 'one-sample/one-measurement' approach to bio-material evaluation by high-throughput and combinatorial chemistry strategies. In con-junction with polymer microarray platforms, these have streamlined the process of material discovery, synthesis and screening for skeletal tissue regeneration applications including mesenchymal stem cell (MSC) isolation, growth and differentiation. In general, high-throughput screening of materials is expected to improve the efficiency of materials development, simply by providing the possibility of including more variations in the material properties and of combining these variations in order to build the optimal material for intended application. In addition, the application of a high-throughput strategy for evaluation of materials is particularly promising when it comes to developing *in vitro* assays with a higher predictive value for the *in vivo* or even clinical functioning of a bone graft substitute. High-throughput screening of materials for bone graft substitutes is still at a relatively early stage, and in the following sections we will elaborate on our experience with regard to high-throughput screening of materials intended for orthopaedic applications.

We have examined the efficacy of a large number of polymer blends, generated by mixing different combinations and ratios of well characterized polymers including binary and tertiary polymer blend formulations, to support the attachment, growth and osteo-genic differentiation of human skeletal stem cells for skeletal repair. We used a combi-nation of a HT approach for material formulation along with a microarray platform to analyse over 135 binary polymer blends and an array of tertiary polymer blends for their ability to function as biocompatible matrices for a variety of human skeletal cell populations. The binary mixture of poly(L-lactic acid) and poly(ε-caprolactone) (blended in a ratio of 20% PLLA to 80%/PCL) exhibited a high binding affinity for STRO-1+ skeletal stem cells, as identified by microarray screening, and also demonstrated a remarkable bone-like 3D architecture when examined by scanning electron microscopy (SEM). To examine their bone regenerative potential, 3D scaffolds of polymer blends permissive for skeletal cell attachment were seeded with human bone marrow STRO-1+ skeletal stem cells. In the murine femoral segmental defects, STRO-1+ skeletal stem cells seeded on the PLLA + PCL polymer blend scaffolds demonstrated significant bone regeneration. This approach demonstrates that multidisciplinary approaches which integrate HT polymer blend and microarray analysis for *de novo* tissue formation using skeletal progenitor and stem cell populations have clear potential to improve the quality of life for many, and emphasizes the need for interdisciplinary strategies to achieve significant therapeutic outcome.

9.2.1 Application of high-throughput combinatorial polyurethane libraries for isolation of MSCs

Conventional approaches such as fluorescently activated cell sorting (FACS) and magnetic-activated cell sorting (MACS) for isolation of desired cell types are often time-intensive, and require expensive instrumentation and trained personnel. In an alternative approach, a library of 120 comprehensively characterized polyurethanes (PUs) were contact-printed in a microarray format on an agarose-coated standard microscope slide, and screened for the identification of PUs capable of bone marrow-derived MSC attachment and immobilization (32). This approach allows the effective and rapid generation of detailed structure–function relationships since the high multiplexing ability of the microarray enables functional screening of all 120 PUs for MSC attachment and immobilization under identical conditions in a single experiment.

The approach adopted a simple yet effective protocol involving overnight incubation of the MSC-enriched STRO-1+ population of human bone marrow on the PU microarrays and identified STRO-1+ cell binding to 31 PUs from the library of 120 PUs. These 31 PUs could be categorized as showing low, moderate and high-affinity on the basis of 2–7, 15–25 and 80–120 STRO-1+ cells bound to individual PU spots of the microarray respectively. From the complete library of 120 PUs, only four, namely PU-16, PU-17, PU-61 and PU-71, were identified as high-affinity PUs and determined to be highly selective for the STRO-1+ population of adult human bone marrow (incapable of binding to the STRO-1+ fraction of human foetal skeletal cells and STRO-1 immunoreactive early osteoblast-like human MG63 cells). Furthermore, the four high-affinity PUs were able to selectively immobilize STRO-1 immunolabelled cells from a heterogeneous population of freshly isolated bone marrow-derived mononuclear cells, thereby demonstrating the significant methodological implications of applying these PU substrates in the development of facile strategies for the isolation of MSCs from human bone marrow.

9.2.2 Application of high-throughput polymer blend libraries for augmentation of MSC growth and differentiation

Polymer blending refers to the judicious mixing of two or more polymers to yield new composite materials (binary blends, ternary blends etc.) with unique properties and functionalities that are often not accessible from the individual polymeric compounds themselves. In an attempt to develop distinctive polymeric biomaterials for skeletal tissue engineering applications, a high-throughput approach was used to generate 135 binary polymer blends by mixing different combinations and ratios of seven commercially available, inexpensive and comprehensively characterized polymers, namely polyethylenimine (PEI), chitosan (CS), poly(L-lactic acid) (PLLA), poly(caprolactone) (PCL), poly(ethylene oxide) (PEO), poly(vinyl acetate) (PVAc) and pHEMA.(33)

The 135 pre-mixed binary blends along with the seven individual polymers were contact-printed onto agarose-coated standard microscope slides in a microarray format for rapid identification of composite materials capable of supporting attachment, growth, viability and differentiation of a variety of human skeletal cell populations including primary MSCs. Following overnight incubation with human bone marrow-derived STRO-1+ MSCs on the microarray slides, total number of cells immobilized onto each polymer spot was determined using a high-content imager (Pathfinder software) and the data was analysed to identify STRO-1 MSC-compatible polymer blends. The analysis demonstrated immobilization and attachment of STRO-1+ cells on microarray spots of PLLA/PCL, PEI/PEO, PEI/PHEMA, PEI/PVAc and CS/PEI binary polymer blends, from which the PLLA/PCL binary blend was selected as the 'hit' composite material for further studies since it exhibited distinctive characteristics conducive for MSC attachment, growth and skeletal tissue engineering.

The polymer blend integrates properties of two polymers, PLLA and PCL, which have FDA approval for clinical application. Although PLLA and PCL individually displayed negligible STRO-1+ MSC attachment, the cells were able to bind to most microarray spots composed of PLLA and PCL blended in different proportions/ratios (i.e. 10:90, 20:80, 30:70, 50:50, 60:40). In particular, a significantly high number of cells were bound to microarray spots of PLLA and PCL blended in a ratio of 20:80. Moreover, scanning electron micrographs of the PLLA/PCL (20/80) blend revealed a highly porous and interconnected 3D architecture that was comparable to natural trabecular bone porosity. Therefore, 3D scaffolds of the PLLA/PCL (20/80) binary blend were fabricated to assess whether the composite material supported robust MSC growth and osteogenic differentiation *in vitro* and *in vivo* and hence was suitable for use in bone regeneration strategies.

Examination of long-term *in vitro* cultures of STRO-1+ MSCs on 3-D PLLA/PCL (20/80) polymer blend scaffolds indicated that the binary blend functioned as an excellent biomimetic template that was able to promote MSC attachment, viability and expression of alkaline phosphatase, an early marker of osteoblast differentiation. Furthermore, PLLA/PCL (20/80) scaffolds supported the generation of mature osteoblasts from the multipotent STRO-1+ MSC population following a 28-day culture period in presence of osteogenic growth factors such as BMP-2 and dexamethasone. *In vivo*, the PLLA/PCL (20/80) scaffolds were able to support the generation of new bone/osteoid to restore the lost bone tissue in critical-size defects created in femora of immunodeficient mice. Although bone regeneration was observed in defects containing PLLA/PCL (20/80) scaffolds alone and scaffolds seeded with MSCs, indices of bone histomorphometry, namely bone volume, bone volume/total volume, trabecular number and trabecular spacing, routinely applied to assess new bone generation, were significantly improved in defects containing MSC-seeded scaffolds compared with scaffolds without cells. The *in vitro* and *in vivo* studies therefore demonstrate the suitability of the PLLA/PCL (20/80) scaffold as a temporal substrate supporting the activities of MSCs including viability, migration, proliferation and comprehensive osteogenic differentiation that are fundamental for bioengineering robust skeletal tissues.

Box 9.2 Classic experiment

Development of Bioglass for bone regeneration

In 1967, Larry Hench made the seminal discovery of Bioglass, the first synthetic material able to bond to bone. He was then assistant professor at the University of Florida, working on semiconducting glass ceramics resistant to high energy radiation for satellite electrical systems. On his way to present his findings at the US Army Materials Research Conference in Sagamore, he made the acquaintance of a Colonel Klinker, who had just returned from Vietnam. After giving a brief description of his work to the colonel, Hench was surprised to find himself challenged by Klinker: 'If you can make a material that will survive exposure to high energy radiation, can you make a material that will survive exposure to the human body?' The colonel depicted to Hench the great number of limb amputations performed on the war casualties, and how the injured bodies systematically rejected the metallic and plastic implants – the only materials available at that time – used by the surgeons to repair the limb.

  In 1968, Hench submitted a proposal to the US Army Medical Research and Development Command, with the following hypothesis: 'The human body rejects metallic and synthetic polymeric materials by forming scar tissue because living tissues are not composed of such materials. Bone contains a hydrated calcium phosphate component, hydroxyapatite [HA] and therefore if a material is able to form a HA layer *in vivo* it may not be rejected by the body.' The project was funded; Hench started to develop a glass containing calcium and phosphate oxide. He provided Dr Ted Greenlee with rectangular samples of his first glass for *in vivo* testing in a rat femoral model. After six weeks, Greenlee called: 'Larry, what are those samples you gave me? They will not come out of the bone. I have pulled on them, I have pushed on them, I have cracked the bone and they are still bonded.' The 45S5 composition of the first Bioglass has since been extensively studied to understand the bioactive properties of the material. The bone-bonding ability of the Bioglass has been linked to the dissolution/reprecipitation phenomenon occurring at the surface of the material in the presence of body fluids. To date, bioactive glasses are still evolving and are used in clinical procedures such as maxillofacial reconstruction (2, 34).

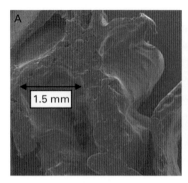 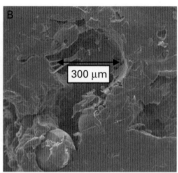 

Figure 9.3   A and B, SEM demonstrates pores of between 0.3 and 1.5 mm within the PLLA/PCL scaffold. C, Polymer blend scaffold cylinder prepared for implantation.

9.2.3   Identification, fabrication and testing of candidate polymers for large animal studies

The single most promising binary blend polymer (PLLA/PCL 20/80), as defined by appropriate physical as well as biological properties, was selected for large-scale fabrication and evaluation within a large animal study. Polymers were fabricated into 18 × 35 mm cylinders by chloroform dissolution and blending, followed by freeze-drying. In order to maximize cell penetration, a central 8 mm diameter medulla was drilled and the surface layer and polymer film removed. The overall size was chosen to replicate the approximate requirements in human clinical practice. Initial large-scale cell permeability tests were performed using negative pressure stain penetration analysis and electron microscopy, to ensure adequate porosity and pore interconnectivity (Figure 9.3).

9.2.4   Identification of the appropriate large animal model

Practical obstacles to *in vivo* scale-up must be addressed to reliably assess new strategies for skeletal tissue engineering before pre-clinical studies can take place. In addition to the requirement to manufacture larger biomaterials and the need to culture cells on a much greater scale than is routine in most research laboratories, further challenges include the establishment of appropriate large animal models for the clinical scenarios to be modelled (35). Although the bone composition of the dog, sheep, goat and pig is similar to those of humans, the age and rate at which osteonal remodelling occurs varies considerably. Additional key factors include cost, local availability, temperament and husbandry expertise (35, 36). For skeletal regeneration models, sheep and goats are particularly relevant as their body weight is comparable to humans and the dimensions of their long bones allow for the use of human implants and fixation techniques (36–38). In the current study, we refined a segmental critical-sized tibia defect model in adult Northern Mule sheep, using an external fixator for stabilization. External fixation is a common clinical technique for human fractures in long bones and variants are also used in distraction osteosynthesis and bone transport techniques for bone regeneration (39). so this model was felt to be most appropriate to the human scenario.

9.2.5   Experimental and surgical technique using the candidate polymer

Following cadaveric pilot studies to refine techniques and confirm an adequate projected sample size, 5 ml iliac crest bone marrow was harvested from each of 12 sheep under general anaesthetic and with full aseptic precautions, and cultured in osteogenic conditions (basal culture medium supplemented with ascorbate and dexamethasone). Sheep were divided into three groups, each containing four sheep: group 1 – empty defect (negative control); group 2 – unseeded scaffold alone (positive control), and group 3 – scaffold seeded with autogenous bone marrow cells (treatment group). Scaffolds were incubated *in vitro* in the absence or presence of 5 × 10 cells/ml in osteogenic medium for 7 days prior to transfer into the *in vivo* ovine segmental defect. After pre-medication, antibiotic prophylaxis, induction and general anaesthetic, a unilateral 3.5 cm tibia defect was created in each sheep, and a scaffold inserted for groups 2 and 3 only (Figure 9.4).

A custom-made external fixator (Orthofix, Maidenhead, UK) was applied using three bicortical Schanz pins on each side of the defect. The sheep were allowed to weight-bear fully with no additional restrictions and radiographs were made immediately post-operatively and at 2, 6 and 12 weeks, at which point the animals were killed and each entire tibia was disarticulated for subsequent computed tomography scans, histological and mechanical analysis (Figures 9.5 and 9.6).

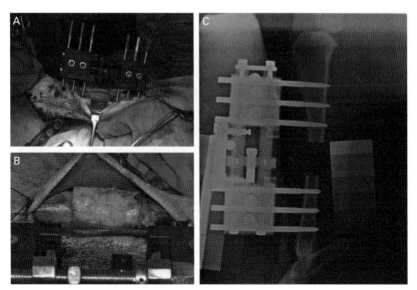

Figure 9.4   A, An empty tibia defect with external fixator *in situ*; B, tibia defect with PLLA/PCL scaffold seeded with cells *in situ*, C, lateral radiograph made immediately post-operatively, demonstrating a 3.5 cm tibia diaphyseal defect and stable construct.

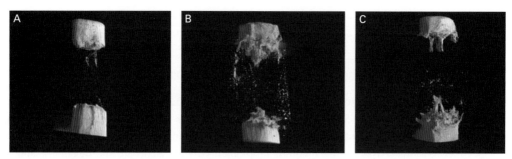

Figure 9.5   Representative CT reconstructions of the defect site after 12 weeks *in vivo* ovine incubation with (A) empty defect; (B) polymer scaffold alone; (C) polymer scaffold and cells. In (B) and (C) there is evidence of bone formation within the pores of the scaffold, and new bone appears to track through the central canal of the scaffold from both proximal and distal bone ends.

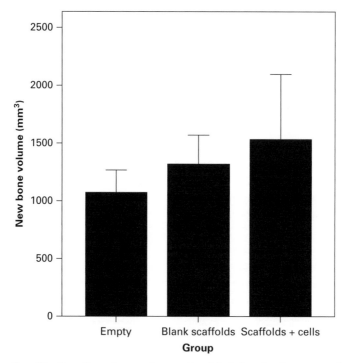

Figure 9.6   Quantification of new bone volume at the tibia defect site shows a trend towards increasing bone healing with the scaffold and with cell-seeded scaffold blend.

Despite encouraging results in preliminary *in vitro* and in small-animal *in vivo* studies, the use of the PPLA/PCL polymer blend failed to significantly enhance new bone formation in the large-animal model as demonstrated by radiological, histological and mechanical analysis. Potential reasons for the failure of this construct in the large-scale tests are myriad, including biological, mechanical, surgical, chemical and analytical factors, all of which are subject to significant variation when upscaling from small animal work and often lead to a disparity between initial encouraging results in laboratory testing and the poorer outcomes of trials in larger species. However, this does highlight the caution that should be exercised before attributing and extrapolating encouraging findings in laboratory tests to the clinical scenario, particularly when implanting heterogenic cells from one species to another and the importance of predictive assays (see Chapter 5 on assay development). By necessity, many independent factors vary simultaneously when upscaling to a large-animal study, and these must be individually defined and addressed before successful clinical translation can occur. Rather, we present here our approach as an example of a continuous facile process through which candidate biomaterials can be identified, fabricated, screened and tested for a variety of clinically useful attributes prior to potential clinical application. The stage is now set to use a similar approach after careful attention to the vascularization of this construct and to cell seeding, and for the analysis of many alternative biomaterial options for tissue regeneration strategies. The pivotal importance of such processes in the development of new biomaterials is clear: putative scaffolds can progress to preclinical evaluation only after confirmation of safety and replicable biological and structural efficacy in a large-animal model.

9.3   Future perspectives

So far, we have attempted to give an overview of classic approaches towards the development of materials for regenerative medicine and to describe the effects that HT screening of materials may have on this process. It is obvious that screening large numbers of materials in the beginning of the process significantly increases the chances of finding positive hits, but also of excluding poor candidates at the early stages of research. Currently, libraries of materials are usually made by varying a single property of the material (e.g. polymer chemistry). However, as described above, not a single property but different properties of a material, including its chemistry, physical and structural properties at different length scales, are responsible for its final performance in the human body. Furthermore, clinicians' requirements regarding handling properties should be taken into account at all times. Therefore, from the clinical perspective, the future of HT screening of materials lies in systems in which different materials properties are simultaneously varied in a systematic way. It is envisaged that arrays of materials are produced that combine physico-chemical and structural properties, to create materials that closely resemble the final implant, but in a less random way than is the case by taking the classical approach towards materials development. Undoubtedly, such systems will require a much more complex library design, assay development and data analysis, but when functional, they will further improve the efficiency of the development of materials for regenerative medicine.

Another envisioned development in the field of HT screening of materials, which is expected to make an enormous impact from the clinical perspective, is coupling of patient diagnostics to materials screens. This offers an approach towards customized healthcare. It is plausible that ideal treatment of a bone defect caused by trauma in a young, active patient may be different from the ideal treatment of a defect created by tumour dissection in an older patient suffering from osteoporosis. If this information on the health condition of a specific patient can be used as an input parameter for the material screen, the success rate of the treatments based on these materials is expected to increase tremendously. Again, coupling patient information to materials screen is far from trivial, and the question remains whether such coupling is possible at all without compromising the reliability of the screens or making them too complex, but materiomics certainly offers larger chances of developing such systems in the future, than classic implant development approaches do.

9.4   Snapshot summary

- An ageing population and increased expectation from patients to retain a high quality of life drive a growing need for regenerative medicine approaches to regenerate damaged and diseased organs and tissues.
- Synthetic biomaterials present an interesting alternative to a patient's own tissue in regenerative medicine owing to their off-the-shelf availability in large quantities, relative ease of production and low costs.
- The classic candidate approach towards development of materials for regenerative medicine includes rational design that often originates in mimicking of the physico-chemical, (surface) structural and mechanical properties of natural tissue.
- Upon design and development, a biomaterial needs to be exposed to a number of pre-clinical *in vitro* and *in vivo* tests aimed to demonstrate its safety and functionality.
- In addition to the process of pre-clinical tests related to the research and development phase, specific regulatory requirements need to be met in order to bring a medical device based on a synthetic material to the clinic.
- High-throughput evaluation of biomaterials for regenerative medicine in the early stages of development is expected to increase the efficiency as well as the success rate of this process.
- In order to make HT evaluation of materials the standard approach for biomaterial research and development, it will be critical to develop reliable assays that are predictive of clinical performance.
- In the future, material libraries will become more complex, and will allow systematic evaluation of various properties of the materials in a single screen.
- From the clinical perspective, future developments will need to focus on combining material screens with patient diagnostics in order to develop customized regenerative approaches.

Further reading

Ratner BD, Bryant SJ. Biomaterials: where we have been and where we are going. Annu Rev Biomed Eng. 2004;**6**:41–75.

Huebsch N, Mooney DJ. Inspiration and application in the evolution of biomaterial. Nature. 2009;**462**(7272):426–32.

Hook AL, Anderson DG, Langer R *et al*. High-throughput methods applied in biomaterials development and discovery. Biomaterials. 2010;**31**(2):187–98.

Khan F, Tare RS, Kanczler JM, Oreffo ROC, Bradley M. Strategies for cell manipulation and skeletal tissue engineering using high-throughput polymer blend formulation and microarray techniques. Biomaterials. 2010 Mar;**31**(8):2216–28.

References

1. Williams DF. *The Williams Dictionary of Biomaterials*. Liverpool University; 1999.
2. Ratner BD. A history of biomaterials. In: Ratner B, Hoffman AS, Schoen FJ, Lemons JE, eds. *Biomaterials Science: An Introduction to Materials in Medicine*. Elsevier; 2004. p. 10–19.
3. Apple DJ. Sir Harold Ridley and his fight for sight: he changed the world so that we may better see it. SLACK Incorporated; 2006.
4. Davison JA, Kleinmann G, Apple DJ. Intraocular lenses. In: Tasman W, Jaeger EA, eds. *Duane's Clinical Ophthalmology*. Lippincott Williams & Wilkins; 2006.
5. Vongpatanasin W, Hillis LD, Lange RA. Prosthetic heart valves. New Engl J Med. 1996;**335**(6):407–16.
6. Hoerstrup SP, Sodian R, Daebritz S *et al*. Functional living trileaflet heart valves grown in vitro. Circulation. 2000;**102**(90003):III–44–49.
7. Damien CJ, Parsons JR. Bone graft and bone graft substitutes: a review of current technology and applications. J Appl Biomater. 1991 Jan;**2**(3):187–208.
8. Dalton P, Harvey A, Oudega M, Plant G. Tissue engineering of the nervous system. In: van Blitterswijk CA, ed. *Tissue Engineering*. Academic Press; 2008. p. 611–47.
9. Padera Jr RF, Schoen FJ. Cardiovascular medical devices. In: Ratner B, Hoffman AS, Schoen FJ, Lemons JE, eds. *Biomaterials Science: An Introduction to Materials in Medicine*. Elsevier Academic Press; 2004. p. 470–94.
10. Fawcett DW. Bone. In: Bloom W, Fawcett DW, eds. *A Textbook of Histology*. W.B. Saunders Company; 1986. p. 199–238.
11. Marks SC, Hermey DC. The structure and development of bone. In: Bilezikian JP, Raisz LG, Rodan GA, eds. *Principles of Bone Biology*. Academic Press; 1996. p. 3–14.
12. Derkx P, Nigg AL, Bosman FT *et al*. Immunolocalization and quantification of noncollagenous bone matrix proteins in methylmethacrylate-embedded adult human bone in combination with histomorphometry. Bone. 1998;**22**(4):367–73.
13. van Gaalen S, Kruyt M, Meijer G *et al*. Tissue engineering of bone. In: van Blitterswijk CA, ed. *Tissue Engineering*. Academic Press; 2008. p. 560–610.
14. Davies JE, Hosseini MM. Histodynamic of endosseous wound healing. In: Davies JE, ed. *Bone Engineering*. Em squared Inc.; 2000. p. 1–14.
15. Friedenstein AY. Induction of bone tissue by transitional epithelium. Clin Orthopaed Related Res. 1968;**59**:21–37.
16. Barradas AM, Yuan H, van Blitterswijk CA, Habibovic P. Osteoinductive biomaterials: current knowledge of properties, experimental models and biological mechanisms. Eur Cells Mater. 2011;**21**:407–29.

17. Karageorgiou V, Kaplan D. Porosity of 3D biomaterial scaffolds and osteogenesis. Biomaterials. 2005;**26**(27):5474–91.
18. Habibovic P, Yuan H, van Der Valk CM *et al.* 3D microenvironment as essential element for osteoinduction by biomaterials. Biomaterials. 2005;**26**(17):3565–75.
19. Habibovic P, de Groot K. Osteoinductive biomaterials – properties and relevance in bone repair. J Tissue Eng Regen Med. 2007;**1**(1):25–32.
20. Covani U, Giacomelli L, Krajewski A *et al.* Biomaterials for orthopedics: a roughness analysis by atomic force microscopy. J Biomed Mater Res A. 2007 Sep;**82**(3):723–30.
21. Schmitz JP, Hollinger JO, Milam SB. Reconstruction of bone using calcium phosphate bone cements: A critical review. J Oral Maxillofac Surg. 1999;**57**(9):1122–6.
22. Weiss P, Gauthier O, Bouler J-M, Grimandi G, Daculsi G. Injectable bone substitute using a hydrophilic polymer. Bone. 1999;**25**(2):67S–70S.
23. Hou Q, De Bank PA, Shakesheff KM. Injectable scaffolds for tissue regeneration. J Mater Chem. 2004;**14**(13):1915–23.
24. Habibovic P, Woodfield T, de Groot K, van Blitterswijk CA. Predictive value of in vitro and in vivo assays in bone and cartilage repair – what do they really tell us about the clinical performance? Adv Exp Med Biol. 2007;**585**:327–60.
25. Brittberg M, Lindahl A. Tissue engineering of cartilage. In: van Blitterswijk CA, ed. *Tissue Engineering*. Academic Press; 2008. p. 534–57.
26. Karam JP, Muscari C, Montero-Menei CN. Combining adult stem cells and polymeric devices for tissue engineering in infarcted myocardium. Biomaterials. 2012:**33**(23):5683–95.
27. Jansen JA. Animal models for studying soft tissue biocompatibiliy of biomaterials. In: An YH, Friedman RJ, eds. *Animal Models In Orthopaedic Research*. CRC Press; 1999. p. 393–405.
28. An YH, Friedman RJ. Animal models of bone defect repair. In: An YH, Friedman RJ, eds. *Animal Models In Orthopaedic Research*. CRC Press; 1999. p. 241–60.
29. An YH, Friedman RJ eds. *Animal Models In Orthopaedic Research*. CRC Press; 1999.
30. Bianco RW, Grehan JF, Grubbs BC *et al.* Large animal models in cardiac and vascular biomaterials research and testing. In: Ratner B, Hoffman AS, Schoen FJ, Lemons JE, eds. *Biomaterials Science: An Introduction to Materials in Medicine*. Elsevier Academic Press; 2004. p. 379–96.
31. de Vries RBM, Buma P, Leenaars M, Ritskes-Hoitinga M, Gordijn B. Reducing the number of laboratory animals used in tissue engineering research by restricting the variety of animal models. Articular cartilage tissue engineering as a case study. Tissue Eng Part B Rev. 2012;**18**(6):427–35.
32. Tare RS, Khan F, Tourniaire G *et al.* A microarray approach to the identification of polyurethanes for the isolation of human skeletal progenitor cells and augmentation of skeletal cell growth. Biomaterials. 2009;**30**(6):1045–55.
33. Khan F, Tare RS, Kanczler JM, Oreffo ROC, Bradley M. Strategies for cell manipulation and skeletal tissue engineering using high-throughput polymer blend formulation and microarray techniques. Biomaterials. 2010;**31**(8):2216–28.
34. Hench LL. The story of Bioglass. J Mater Sci. Materials in medicine. 2006;**17**(11):967–78.
35. Reichert JC, Saifzadeh S, Wullschleger ME *et al.* The challenge of establishing preclinical models for segmental bone defect research. Biomaterials. 2009;**30**(12):2149–63.
36. Muschler GF, Raut VP, Patterson TE, Wenke JC, Hollinger JO. The design and use of animal models for translational research in bone tissue engineering and regenerative medicine. Tissue Eng B Rev. 2010 9;**16**(1):123–45.

37. Newman E, Turner AS, Wark JD. The potential of sheep for the study of osteopenia: Current status and comparison with other animal models. Bone. 1995;**16**(4):S277–S284.

38. Reichert JC, Epari DR, Wullschleger ME *et al.* Establishment of a preclinical ovine model for tibial segmental bone defect repair by applying bone tissue engineering strategies. Tissue Eng B, Rev. 2010;**16**(1):93–104.

39. Smith JO, Aarvold A, Tayton ER, Dunlop DG, Oreffo ROC. Skeletal tissue regeneration: current approaches, challenges, and novel reconstructive strategies for an aging population. Tissue Eng. B Rev. 2011;**17**(5):307–20.

10 Non-biomedical applications of materiomics

Carson Meredith, Sangil Han, Ismael Gomez,
Johannes Leisen and Haskell Beckham

Scope

Materiomics approaches may be applied broadly in the design and characterization of a wide range of materials, going beyond the biomedical focus. This chapter presents essential concepts relevant to the implementation of materiomics to physical and chemical properties not usually associated with biomedical materials. Recent progress in this area is reviewed briefly with several examples given in high-throughput measurement of organic and inorganic materials properties. The applications that are covered range from speciality coatings to membranes for fuel cells and metal–organic framework materials for carbon capture. Properties treated here include mechanical, spectroscopic and transport characteristics.

10.1 Tutorial on basic principles

10.1.1 Materials complexity

While previous chapters have focused on the application of materiomics concepts to biomedical materials, this chapter will emphasize how and why materiomics is being employed in materials research and development beyond the biomaterials field. The central challenge inherent to materials research and development is that most products consist of *multiple components* whose final properties are usually a sensitive function of *composition* and *processing conditions*. The complex interactions make it difficult, if not impossible, to design products *a priori* without significant experimentation. Academicians are often interested in characterizing fundamental phenomena, such as phase transitions or magnetization, and discovering how these phenomena can be used to develop novel materials functionalities. Industrial researchers are often focused more on discovery of chemistry and structure of novel materials, as well as their optimization by adjusting composition and processing conditions, in order to produce a new or improved product. However, in both the fundamental and applied cases, the problem of complexity between chemistry, structure, processing and properties is the central challenge, and often is a limitation in discovery of new knowledge or materials.

Materiomics: High-Throughput Screening of Biomaterial Properties, ed. Jan de Boer and Clemens van Blitterswijk. Published by Cambridge University Press. © Cambridge University Press 2013.

The characterization of materials phenomena, and the discovery of new materials, underlies a large number of applications critical to society and industry. In addition to biomaterials, applications in which materials development is a critical part of solutions include:

- fuel cells and solar cells for more efficient power generation;
- capture of carbon dioxide ('carbon capture') for reducing greenhouse gas emissions;
- new catalysts for biofuel production or conversion of biomass into chemicals;
- hydrogen storage materials to reduce fossil fuel use;
- microelectronics and organic electronics for computers and displays;
- lightweight, high-strength materials and adhesives for increasing fuel efficiency in cars and airplanes.

Materials scientists generally classify materials into categories according to material chemistry (metals, ceramics, organic/polymeric, and hybrids of these), structure (crystalline versus amorphous, heterogeneous versus homogeneous), or function (insulating, semiconducting, conducting, for example). Examples of complexity abound in each materials class, and in this chapter we will give examples of how materiomics approaches are being applied to overcome these challenges in each class of materials. Many final products are mixtures of these various classes, further complicating the picture. Take for example a common product, latex paint, which illustrates very clearly the magnitude and problem of complexity. Paint is primarily an emulsion of acrylic monomer in water, which also contains dispersed inorganic pigment and a host of other additives, including initiator, antioxidants and ultraviolet absorbers (to stabilize against environmental degradation), emulsifiers (to suspend the various particles), wetting agents (so paint wets the brush or roller but also transfers to a surface), and viscosity modifiers (to aid in application). This complex mixture, often exceeding 20 ingredients, is applied as a liquid, but as it dries the monomer polymerizes to form a solid film that binds the coating to a substrate. The dried film microstructure consists of latex particles, inorganic solids and microscopic air voids, along with other organic additives. The microstructure is critical, as it is responsible for the light-scattering properties that provide opacity and influence the colour, appearance and permeability. While fundamental aspects of the structure and function of individual ingredients or simple model structures are well described and can be predicted, the interacting effects of these components, resulting in a microstructure that produces an array of desired final properties, is largely not predictable from first principles. Hence, materials for paints have relied heavily on knowledge gained through trial-and-error experimentation. The large number of components, their potential interactions and the associated phenomena of particle aggregation, film formation, light scattering and absorption, and solvent evaporation lead to a nearly intractable experimental problem. Fortunately for a commodity like paint, decades of knowledge allow useful advances to be made using modest experimental efforts. However, this is not the case for some of the advanced materials listed as current applications above, for which such a large previous knowledge base is usually not available.

Equally important as the choice of *materials chemistry* (e.g., aluminium versus iron) is the effect of *processing*. The formation of specific nano- and microstructures, encompassing crystallinity, foams, dispersions, emulsions and phase-separated microstructures, is often critical to achieving specific functional properties. These structures are usually produced during the *processing steps* by some combination of thermodynamically driven phase-separation and/or reactions, which are usually controlled kinetically during processing. Often, this involves the timed addition of catalyst or initiator, heating to elevated temperature, application of mixing, and controlled quenching back to room temperature. Just as with composition effects, our ability to predict accurately, *a priori*, the interacting effects of these types of processing steps is quite limited. In any reasonably complex system, experiments are required. However, the number of experiments required to 'map out' the combined effects of chemistry and processing changes for multicomponent materials is vast and easily overwhelms the capabilities. Hence, materials research is in general a very labour-intensive and time-consuming task and that is complicated by the interdependence of chemistry, structure and processing described above. In the past 15–20 years, developments in each of these areas have set the stage for the emergence of materiomics as a viable approach in a wide range of materials research and development problems, to be described in this chapter. Some recent examples include screening of thin polymer coatings (1), photoluminescent materials (2), fire-retardant plastics for aircraft (3), development of materials for hydrogen storage, (4) and discovery of novel electrode materials for fuel cells (5) and lithium-ion batteries (6).

10.1.2 Materials properties

While an enormous number of materials characterization tests and measurements exist, a number of categories of properties repeatedly show up in a wide variety of applications and investigations. This section briefly reviews those properties, and discusses why they present challenges to materiomics.

Microstructure and phase behaviour
The characterization of heterogeneous phases (crystalline regions, for example), including their composition, size, shape and interconnectivity, is of paramount importance in nearly all areas of materials development. Optical and fluorescence microscopy techniques, while limited in resolution to ~100 nm, are readily automated. Higher-resolution electron microscopy and scanning probe methods are somewhat less amenable to automation, although examples do exist. Fourier transform infrared (FTIR) spectroscopy and X-ray scattering methods, also readily applied to materiomics, are useful for identifying the presence of amorphous versus crystalline materials, and for X-rays in particular the size and crystalline form can be identified.

Mechanical properties
In many materials problems, such as structural components or films, the mechanical response to deformation is a critical consideration. Properties of interest include, but are not limited to, tensile strength, elongation at break, elastic modulus, yield strength

and impact strength. A challenge inherent to applying these tests in a high-throughput mode includes the usually destructive nature of the assay, requirement of physical contact with the sample, and geometry and rate dependence of measured properties. Many industries rely upon standardized tests (ISO, ASTM) that were developed decades ago, before the use of high-throughput materiomics became widespread. Hence, adaptation of mechanical property testing to materiomics is a challenge for industrial acceptance, although some examples do exist. Materiomics has been successfully applied to develop scratch- and abrasion-resistant coatings used in the automotive and aircraft industries, for example (7).

Electronic and optical properties

The ability to transport electrons is a key consideration in applications such as electronics and semiconductors. Properties of interest include ac or dc conductivity, ac impedance, semiconductor electron or hole mobilities, as well as photoelectric effects for solar cells. Optical materials include light emitting diodes and luminescent materials for displays and sensors, in which properties such as fluorescence, emission, absorption and scattering effects for ultraviolet, visible and near-infrared light for microelectronics are relevant. While many optical and spectroscopic properties are readily measured in high-throughput mode because automated instrumentation is available, materiomics application to conductivity and impedance measurements is more challenging and usually requires customized instrumentation.

Transport, sorption and adsorption properties

In applications where a chemical species dissolves in a material, or adsorbs on its surface, it is important to be able to measure the sorbed or adsorbed amounts and capacities, and for micro- or mesoporous materials the surface area is important. When the rate of transport of such chemical species through a film or surface is also important, then properties such as diffusion and permeability are important. The applications in which these properties are of interest include barrier films for packaging, as well as adsorbents for hydrogen storage and CO_2 capture. A challenge is the need to expose to external penetrants for long periods of time, and often measurement of two penetrant species is required to ascertain selectivity, often represented as the relative permeabilities of two gases, e.g. O_2 and N_2. Conventional instrumentation for these types of measurements is adapted for high accuracy under steady state conditions, and is often not ideal for the rapid serial or parallel measurements called for in materiomics. Later, we will discuss some recent work in development of high-throughput instrumentation for sorption and diffusion of gases in novel CO_2 adsorbents.

Reaction properties

Nearly 90% of industrial chemical processes are based on catalysis, which involves use of a material (catalyst) that increases reaction rate and/or selectivity towards desired products (8). Applications of catalysis include chemicals production, petroleum refining, air and water purification and hydrogen production, to name a few. However, selection of catalysts with appropriate reaction kinetics and selectivity, and screening their sensitivity to reaction conditions (temperature, pressure, composition of feed, presence of impurities) is extremely challenging with conventional one-sample-per-experiment approaches. Significant progress has been made in catalyst discovery via materiomics approaches, including catalysts for novel speciality plastics (9), hydrogen production from methanol (10), and fuel cell electrodes (5).

Informatics challenge

Because most materials measurements have traditionally been carried out 'one-sample-at-a-time', materials research has usually not required data mining and informatics strategies. However, combinatorial high-throughput (CHT) strategies generate unusually large amounts of data in even a single experiment. In order to discover patterns within large datasets that contain complex interdependencies, efficient informatics approaches tailored to materials science must be employed. Generally, informatics involves a comprehensive approach to managing CHT data, including experimental planning, databases for data storage, instrument control, processing of data, and finally data mining (pattern recognition and predictive modelling) (11). These challenges need to be met in order to incorporate materiomics fully into a research and development workflow. Advances in informatics for materiomics have been made particularly in discovery and development of novel catalysts for industrial chemical reactions, which will be reviewed below.

10.1.3   Combinatorial synthesis and high-throughput methods

To address the challenges above, recent decades have witnessed the development of combinatorial and high-throughput screening (CHTS) methods to improve the efficiency of materials discovery and characterization. A number of reviews and monographs have been published on the subject of materials CHTS, featuring applications outside the biomaterials field (11, 12 , 13–24). While intense activity has occurred in the past two decades, one can find examples of accelerated or parallel screening approaches scattered throughout the history of materials and chemistry research. Kennedy *et al.* described the use of composition-gradient samples to explore phase behaviour and superconductivity of inorganic ternary mixtures (25). Hanak and co-workers, in 1970, formally proposed the parallel screening of gradient mixtures as the 'multiple sample concept' and foresaw the adoption of such methods in materials research (26). In pharmaceutical research, the 'rediscovery' of Merrifeld peptide synthesis using a combinatorial library approach in 1991 led to the widespread adoption of these methods in drug discovery (27). However, despite early attempts like these, the application of CHT experimental methods to materials in general did not begin to accelerate significantly until 1995, with the publication of a paper that demonstrated the discovery of new high-critical-temperature (high-T_c) superconductors using arrays of inorganic materials (28). A number of new commercial products have been developed based on CHT discovery and development, including a catalyst for manufacturing speciality propylene–ethylene copolymers (Dow Chemical, Versify plastomers and elastomers) that are useful as components of rigid packaging, roofing materials and plastic tubing. Other commercial successes that began with CHT methods include a phosphor for displays (Agfa-Gevaert NV) and a polymer used in electronic applications (JSR Corporation) (9).

Combinatorial sample libraries contain a large number of conditions, e.g. composition and temperature, on a relatively small sample footprint, e.g. a titre plate, microscope slide or microfluidic chip. To achieve significant increases in exploration rate, the libraries are subjected to HT screening experiments chosen judiciously to detect meaningful properties. In addition to the development of experimental strategies has been the increased recognition of the importance of informatics and data mining techniques to HT materials discovery and characterization (23). The next section will review a number of examples of these developments.

Box 10.1 Classic experiment

Identification of a blue photoluminescent composite material from a combinatorial library

Considerable interest exists in advanced luminescent materials for flat panel displays. Since these properties arise from complex interactions among the host structure, dopants, defects and interfaces, combinatorial materiomics strategies are expected to accelerate identification of new materials. Wang *et al.* reported a new blue photoluminescent (PL) material, $Gd_3Ga_5O_{12}/SiO_2$, identified with combinatorial screening (2). Libraries were prepared by molecular beam epitaxy to deposit thin-film precursors at different sites on a substrate, using a series of patterns produced with photolithography.

The lithographic method allowed precise fixation of independent sample regions at 650 μm by 650 μm, sites spaced 100 μm apart. This resulted in 4 different compositions consisting of n elemental components. Based on knowledge that some silicate- and gallate-based phosphors have band gaps desirable for display applications, the library was prepared with SiO_2 and Ga_2O_3 precursors as host materials deposited on SiO_2. A number of activator layers were grown on the library also, including Gd_2O_3, EuF_3, CeO_2, ZnO and Y_2O_3. Figure 10.1A shows a visible-light image of a typical library, and Figure 10.1B shows PL under a UV lamp. Roughly 25 sites in the library showed substantial blue PL, each with the common precursors Gd_2O_3 and Ga_2O_3. The brightest site, at which only these two precursors were deposited, corresponded to the nominal composition $Gd_{5.2}Ga_{3.33}O_z$. X-ray photoemission spectroscopy confirmed the presence of SiO_2 at the site with optimal PL, which appears to be related to the mechanism of luminescence. Blue PL appeared to arise from nanocrystalline or amorphous SiO_2 or interfacial silicate that was finely dispersed into the $Gd_3Ga_5O_{12}$ matrix, and possibly coated on the surface of $Gd_3Ga_5O_{12}$ grains, which are of the order of 100 nm in size. These interfaces may form the specific local electron states that give rise to the observed blue emission, pointing to a complex origin of the PL emission.

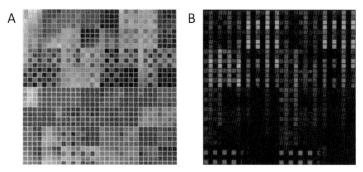

Figure 10.1   A library of novel luminescent molecules generated by epitaxial growth of gallinule, silicate and numerous activators, through a series of physical masks. A, Visible light image; B, photoluminescence under UV excitation. Reproduced with permission from (2).

10.2   Relation to materiomics

10.2.1   Library preparation

Figure 10.2 presents some examples of discrete and gradient combinatorial library preparation strategies. Discrete libraries are convenient, since they can usually be produced using commercially available instrumentation, including molecular beam epitaxy or ion implantation (for solids) (Figure 10.2A) and liquid handling robots. Figure 10.2A shows a schematic of a molecular beam epitaxy apparatus that can be used to prepare libraries with a wide variety of combinations of inorganic compounds, such as phosphorescent, semiconducting, dielectric, magnetic and superconducting materials (29). Two specific examples include two-dimensional libraries of compound semiconductor thin films produced by ion implantation (30) and thin film libraries of epitaxy-deposited (Ba, Sr)TiO$_3$ capacitors for memory and tunable microwave device applications (31). Additional methods to prepare libraries of compounds include micro-fluidic (32) and inkjet techniques (33), which allow small droplets of complex fluids with controlled composition to be prepared as an emulsion or a coating, respectively. Recently, a liquid source misted chemical deposition (LSMCD) method was demonstrated for creating libraries of inorganic materials, with ppm level composition control (34). In contrast to discrete sample preparation, continuous-gradient (Figure 10.2B and 10.2C) libraries are inherently parallel since a single sample actually contains a range of conditions, as recently reviewed (35). Gradient-based libraries can, however, be sectioned if needed for characterization with serial or point-to-point analytical measurements. The use of samples containing continuous gradients in polymer properties was demonstrated by Meredith *et al.* (36–38). Libraries containing the equivalent of several hundred distinct chemistries, microstructures and roughnesses were prepared on samples the size of a microscope slide by using composition spreading (39) (Figure 10.2B), temperature gradient and thickness gradient (37) (Figure 10.2C) techniques. Recently, a 'universal' functional gradient preparation method was demonstrated, by which a variety of species can be attached by 'click' chemistry, e.g. azide-functionalized molecules (40).

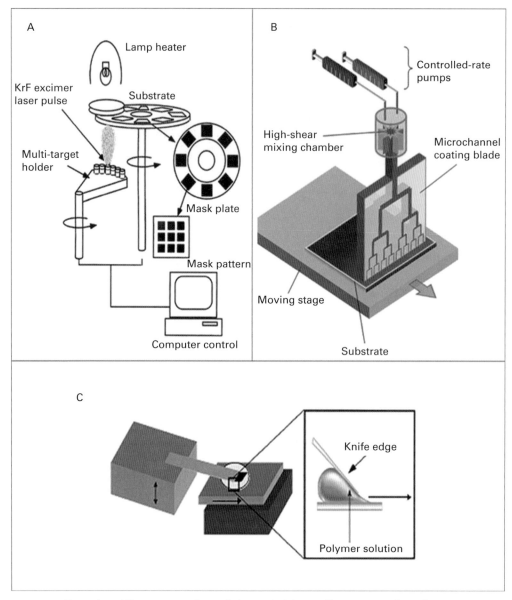

Figure 10.2   Examples of library preparation techniques. A, Discrete library preparation using molecular beam epitaxy through mask arrays, often applied to inorganic materials. Reproduced with permission from (29). Copyright © 2002 Elsevier Masson SAS. All rights reserved. B, Composition-gradient coating process used to prepare polymeric gradient libraries. Reproduced with permission from (39). C, Thickness-gradient coating process used to prepare polymer gradient libraries. Reproduced with permission from (37).

Molecular beam epitaxy methods can also be applied to produce composition gradients of inorganic materials (31). Twin screw extrusion, a common commercial process for blending plastics, can be used to produce gradients in blend composition. This approach has been applied to explore composition effects on mechanical properties in polymer nanocomposites for high-strength materials (41).

10.2.2   Electronic and electrochemical properties

Ion beam implantation has been applied to prepare combinatorial libraries of buried II/VI compound semiconductor nanocrystals, consisting of combinations of different compositions of Cd and Se implanted into thin silica films. These materials are of interest as novel photoluminescent materials for displays and sensors. Figure 10.3A shows an example of one such library, which can be subsequently screened for optical photoluminescence properties (42). Identification of new organic light-emitting device (OLED) materials for displays is a challenge owing to the sensitivity of light-emission efficiency to combinations of several thin layers of conducting and semiconducting materials. Properties are quite sensitive to the purity, composition and thickness of each layer, as well as their interfacial adhesion. The combinatorial fabrication and screening of OLED arrays that allow rapid, efficient screening of layer thickness and chemistry has been described (43). Figure 10.3B shows an image of one such 21×21 OLED array library, where the difference in shade (or in colour, in the original) indicates the sensitivity of optical emission to layer thickness.

Ferroelectric materials, used in diverse applications including capacitors, non-volatile memory and ultrasound imaging, have been developed using combinatorial libraries that explore compositions containing Bi, La, Ce and Ti_3O_{12}, in order to identify materials with high remnant polarization (a measure of the ability to form electronic dipoles under an electric field – the key property for classifying ferroelectric materials) (34). A newly developed scanning probe microscope method was found to allow reliable, rapid determination of the relative ranking of polarization on the thin-film libraries, and allowed

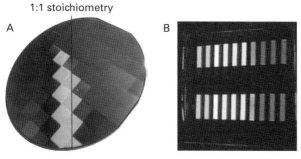

1:1 stoichiometry

A

B

Figure 10.3   A , Cd and Se ion-beam implanted silica thin films in which each square represents a different combination of Cd and Se dose. Reproduced with permission from (30). B, Blue-to-red OLED arrays that explore the effect of thickness combinations in the hole and electron injection layers on optical emission frequency. Reproduced with permission from (43).

identification of an optimal composition with maximum polarization, as well as a new structure–property relationship that correlates crystallinity with polarization.

A number of materials challenges involve electrochemistry, including battery development, electrocatalysis, photocatalysis, corrosion protection, sensor development, photovoltaics and light-emitting materials. Muster *et al.* have reviewed the utilization of high-throughput and combinatorial methods for electrochemical materials (24). One specific example is provided by Jarayaman and Hillier (44), who describe a 'gel-transfer' synthesis of ternary composition gradients of fuel-cell catalyst candidates. The libraries are prepared by diffusion of metal salts into a hydrated gel to establish spatially varying concentration fields, followed by electrodeposition on a substrate. To illustrate the utility of this method, a platinum–ruthenium–rhodium ($Pt_xRu_yRh_z$) catalyst gradient was constructed, and its utility in several fuel-cell reactions was evaluated (44). Zapata *et al.* describe a high-throughput ionic conductivity apparatus that was applied to screen candidate polymer electrolyte membranes (PEMS) for fuel cell applications. The device uses a miniature four-point probe for rapid, automated point-to-point AC electrochemical impedance measurement in both liquid and humid air environments. Screening of 40 novel polyelectrolyte blended membranes was enabled by this approach (45).

10.2.3   Mechanical properties

The measurement of mechanical properties directly on libraries has been accomplished by using miniaturized probes or indenters that deform the library at specific positions, and then record stress and strain. This has been applied using commercially available nano- and microindentation instruments, and atomic force microscopy, as well as customized mechanical screening instruments. For example, mechanical characterization of polymer films has been performed using a high-throughput mechanical characterization (HTMECH) apparatus (46–48). Polymer libraries are deformed at various positions (corresponding to different compositions) using an indenter, and the data is used to extract toughness, strain at break, ultimate tensile strength and elastic modulus. Commercially available instruments, based on a similar concept, but with parallel arrays of mechanical pins, have been reported for the 96-well platform (49). In addition, nanoindentation has been demonstrated as a useful tool for high-throughput screening, as discussed in Chapter 2 (50).

Coatings and thin films used as protective layers in automobiles, aircraft and marine applications, as well as in microelectronics and optical materials, represent a challenge to high-throughput mechanical property measurement. These coatings are inherently complex owing to the interactions between composition, substrate, curing temperature and time, and relative humidity – effects that are difficult to model and predict in combination. Potyrailo and co-workers demonstrated high-throughput screening of abrasion-resistance of silica-silane automotive coatings (7). This technique used an optical screening technique to measure light-scattering induced by surface damage on libraries of coatings, using industrially relevant abrasion standards (7). High-throughput scratch and mar measurements, which are important conventional metrics used for coatings, have also been demonstrated recently for polyurethane (PU) coating formulations. Optical

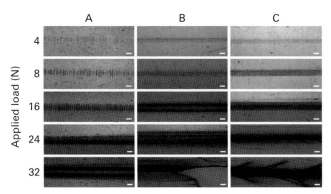

Figure 10.4   Scratch images at various applied loads of poly(urethane) coatings cured using three different temperature–time curing cycles, C (A), C2 (B) and C3 (C). Scale bar indicates 140 μm.

microscope images of scratch deformations made by an automated steel ball-tipped probe are displayed in Figure 10.4. The dependence of scratch depth on loading, composition, curing temperature and time is easily determined through automated image analysis. This screening also indicates the loading at which film failure, characterized by cracking and fracturing, is initiated.

10.2.4   Spectroscopic properties

A wide range of spectroscopic measurements has been applied to screening material libraries, including FTIR, Raman, near-infrared, ultraviolet-visible, fluorescence, mass spectrometry and NMR (11, 51, 52). Serial sampling is usually convenient when a single measurement site is limited to a small 'spot' size (a few millimetres or less), when the measurement probe or detector is prohibitively expensive to reproduce in parallel, and when an existing serial instrument can be easily fitted with automated sample stages (for point-to-point scanning of the library). Parallel screening allows extremely high-throughput rates but is only convenient when the 'field of view' of the detector permits simultaneous inspection of the entire library. For example, digital infrared or fluorescent optical inspection cameras can view an entire library and detect thermal activity (catalyst screening) or presence of a chromophore (materials for displays and sensors). Other examples include characterization of composition in polymer gradient libraries by FTIR microspectroscopy (53) and screening new polymerization catalysts in 96-well micro-reactor arrays using fluorescence spectroscopy (54). Synchrotron X-ray sources have been used to analyse crystalline structure of materials on combinatorial libraries (55). To illustrate how spectroscopic methods are applied in HT mode, we discuss in slightly more detail a novel development in the field of NMR spectroscopy.

Recently, a new approach, termed 1D NMR profiling, was developed for HT characterization of composition-gradient samples (56). As described in detail elsewhere, the 1D NMR profiling technique was demonstrated using a gradient polyurethane library (56), described with the schematic in Figure 10.5. Adjusting the position of the sample in the

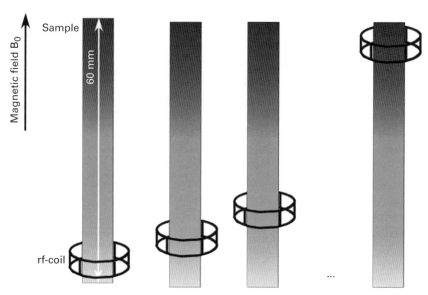

Figure 10.5 Schematic setup of the 1D NMR profiler. The sample is moved through the stationary rf coil in steps that depend on the sensitive field-of-view of the coil. Reprinted (adapted) with permission from (56). Copyright 2012 American Chemical Society.

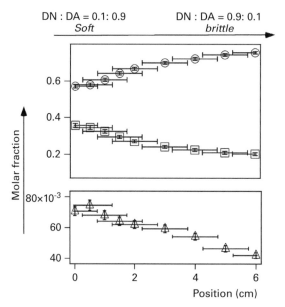

Figure 10.6 Molar fractions of rigid (circle), intermediate (square) and mobile (triangle) components as determined from 1D NMR profiling along a polyurethane library with a gradient in relative isocyanate and polyol content. Reprinted (adapted) with permission from (56). Copyright 2012 American Chemical Society. DN = Desmodur N and DA = Desmodur A (BASF isocyanates).

NMR radiofrequency (rf) coil allows characterization of the gradient library at various positions.

Screening in this manner allows identification of components with various mobilities, e.g. mobile, intermediate and rigid, which are critical indicators of mechanical, barrier and thermal properties (Figure 10.6) (56). At the low isocyanate content end (position marked 0 cm), the film consists of about 57% rigid component, 36% intermediate and 7% mobile component. The rigid-component fraction increases with increasing isocyanate content, while the mobile and intermediate component fractions decrease.

Box 10.2 Classic experiment

Combinatorial study of block copolymer phase behaviour

Although there have been many previous studies of surface pattern formation in block copolymer films, there is no predictive theory of this type of pattern formation and many questions remain about basic phenomenology. For instance, the factors governing the size of the ubiquitous island and hole patterns (e.g. molecular mass M and temperature T dependence) in these films are not well understood. The precise h ranges in which the films remain smooth is another basic film property that requires further investigation. Since these film properties are crucial for the development of a theoretical description of this type of pattern formation, we have combinatorially investigated pattern formation in block copolymer films using a flow-gradient technique to create films having a range of film thickness h and molecular mass M (1). Figure 10.7 shows two thickness-spread libraries of the poly(styrene-b-methyl methacrylate) (PS-PMMA). Near-symmetric poly(styrene-b-methyl methacrylate) (PS-b-PMMA) diblock copolymers with various molecular mass values, M, were used to prepare thickness spreads, which were subsequently annealed. Distinct morphological regions are observed in Figure 10.7, including broad 'bands' where the film surface is smooth and relatively narrow ranges where the film pattern is labyrinthine (we term these 'spinodal patterns' by analogy with phase separation). The regions of hole and island pattern formation (found on either side of the spinodal patterns) are consistent with previous observations on films of fixed thickness, but the observation of broad smooth regions and the spinodal morphology are apparently novel. Smooth films observed for certain h ranges centred about multiples of the lamellar thickness L_0 were attributed to an increase in the surface chain density with h in the outer brushlike copolymer layer. We also observe apparently stable labyrinthine surface patterns for other h ranges, and the average size of these patterns is found to scale approximately as L_0. Hole and island patterns occur for h ranges between those of the labyrinthine patterns and the smooth regions, and their size similarly decreases with L_0 and M.

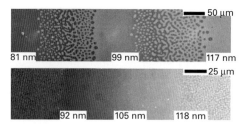

Figure 10.7 Optical micrograph of a molecular weight (MW) 26,000 g mol^{-1} (top) and MW 104,000 g mol^{-1} (bottom) poly(styrene-b-methyl methacrylate) gradient film, showing the addition of lamellae to the surface with increasing thickness. Reprinted with permission from (1). Copyright 2001 American Physical Society.

10.2.5   Transport and adsorption properties

Many applications in chemical processing and separations involve high-surface area, microporous adsorbents or membranes. One example is metal–organic frameworks (MOFs), which are size-selective, high-capacity materials for adsorptive and membrane-based capture of CO_2 and other gases (57). MOFs exhibit nanoporous crystalline structures in which organic linker molecules self-assemble with metal centres to form crystals with well-defined pore size, high surface area and low framework density. Significant problems, most notably the huge design space of metal centres and organic ligands available, frustrate attempts to discover or design such materials. In addition to the large, under-explored design space, some MOFs are unstable in water vapour, and the principles for designing water-stable MOFs are not yet well known. Measurements for determining adsorption isotherms are tedious and time-consuming. For considering practical CO_2 capture applications, the performance of MOFs that have been exposed to water vapour and acid gases like SO_2 and NO_2 is critical, and almost no data addressing this issue are available.

To address the above challenges, several techniques for efficient screening of adsorbent transport and sorption properties have been recently described (57, 58). For example, Han *et al.* describe a high-throughput-sorption instrument that has an array of 36 chambers, each connected to a pressure sensor and a valve (Figure 10.8) (57). By monitoring pressure decay versus time, CO_2 and N_2 adsorption are measured. Eight MOF samples were used to demonstrate the utility of this approach (57). Figure 10.9 shows the observed dry CO_2 uptake for each material. Each sample reaches a different equilibrium pressure with different kinetics. The three materials among the eight tested with the highest CO_2/N_2 adsorption selectivity for dry gases are Co-NIC, Cu-PCN and ZIF-7 (Figure 10.10). No previous information on the adsorption selectivity of Co-NIC or Cu-PCN has been reported.

To probe the robustness of each MOF, each material was exposed in the high-throughput array to humid air at 78% relative humidity (RH) as well as acid gases that are present in flue gas. The equilibrium adsorption properties were measured again after these exposures. The selectivity, an important indicator of relative affinity for CO_2 and

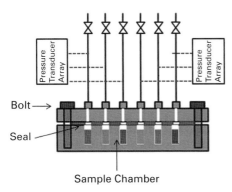

Figure 10.8 Schematic of high-throughput sorption and transport screening system. Reprinted (adapted) with permission from (57). Copyright 2012 American Chemical Society.

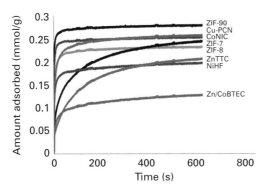

Figure 10.9 Carbon dioxide uptake for eight MOFs before water vapour exposure measured at 30 °C and initial pressure of 14 psi. Equilibrium pressures are different depending on adsorption capacity of the samples. Reprinted (adapted) with permission from (57). Copyright 2012 American Chemical Society.

N_2, before and after these exposures, is given in Figure 10.10. The effects of acid gas exposure, a practical consideration that is difficult to assay with conventional throughput instruments, were considerable in several of the MOFs in Figure 10.10. For most of these compounds, this data represent seminal contributions to the field, especially relative to the effects of water, SO_2 and NO_2 on adsorbent performance.

10.2.6   Informatic approaches

Informatics encompasses the sum of all data management and exploration tools applied to a materiomics workflow, which can include experimental design and planning. The design and planning may consist of *in silico* modelling from first principles, or simply classical design of experiments approaches. Informatics usually is recognized to include coupling experimental design to sample preparation instruments, automation of data acquisition and data mining to identify patterns and anomalies (hits), as well as to

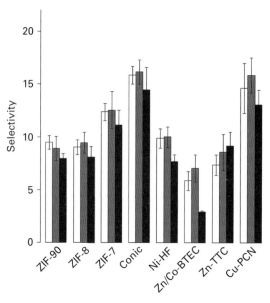

Figure 10.10   Pressure-corrected selectivity of CO_2 to N_2 adsorption using a high-throughput sorption apparatus for eight MOF materials. Measurement before humid air is white, after humid air is grey, and after SO_2 and NO_2 exposure with humid air is black. Reprinted (adapted) with permission from (57). Copyright 2012 American Chemical Society.

discover new structure–property–processing relationships. A thorough overview of experimental design for CHT materials development has been presented elsewhere (59–63). Design of experiments allows researchers to plan the compositions that are explored in a library by using statistical techniques to select combinations most likely to yield useful information. One of the other major challenges inherent to combinatorial datasets is pattern recognition, for which standard visual data inspection is not sufficient owing to the multivariate, complex dependencies. Some standard dimensional reduction techniques have been applied to pattern recognition in CHT screening, including principal components analysis and partial least squares regression. A major application of informatics is in the area of catalyst discovery, where both standard and novel pattern recognition methods have been used to search large datasets of catalyst performance as a function of multiple component compositions (11). For example, Maier and colleagues showed how novel data visualization approaches could be applied to identify the combinations of metals (Cr, Co, Mn, Te and Ni) in a catalyst to achieve optimal activity and selectivity for producing propene oxidation, a common organic reaction (64). Takeuchi discusses data management issues in HT experimentation, and highlights examples where data-mining tools are being implemented for extracting knowledge and predicting new compounds, with an emphasis on electronic materials (65).

Quantitative structure activity relationship (QSAR) analysis has been used in the pharmaceutical industry for decades to describe the biological activity (often protein binding) of candidate drugs as a function of molecular descriptors such as polarizability, atomic structure and lipophilicity (66). However, one of the first applications of QSAR to materials CHT was by Kholodovych *et al.*(60). A semi-empirical model was applied to predict cellular response to the surfaces of a combinatorial library of tyrosine carbonate polymers. Quantitative predictions of cellular response to within 15.8% allowed identification of polymer structural features most associated with variations in the polymer performance, thereby providing direct guidance for the next iterations in design. Laboratory research informatics systems (LRIS) are integrated, commercially available, software packages that streamline data-intensive CHT research strategies. A model open-source LRIS, used at the NIST Combinatorial Method Center, integrates library design, fabrication, measurement and analysis, by incorporating non-proprietary instrument interface and automation software (59–63).

10.3   Future perspectives

The past 10 years have witnessed the emergence of combinatorial and high-throughput screening (CHTS) as a new set of tools in materials development. These tools, developed initially in the drug-discovery industry, and later applied to catalysts, inorganic, and organic polymeric materials, allow orders of magnitude increases in the rate of exploration and characterization of new materials. CHTS promises markedly improved throughput, with modest sacrifices in accuracy, and it appears from the body of work reviewed here that this is achieved in most cases. When speed and efficiency are the only concerns, it is clear that CHTS methods are advantageous over conventional throughput methods. However, another central tenet of CHTS is that enhanced throughput will enable discovery of *new materials* or *new knowledge* in significantly shorter times than by using conventional methods alone. The ultimate 'payoff' for investing in CHTS is the development of new, commercial products or unforeseen advances in knowledge. There are numerous examples where a discovery has been made using CHTS, where the likelihood of making a similar discovery with conventional synthesis in the same amount of time is exceedingly small. However, the hypothesis that CHTS will speed discovery in general remains difficult to test by looking at published work alone. For example, it is not often that researchers carry out the exact same experiments both combinatorially and conventionally under controlled conditions. An important aspect that is often overlooked is the extent of correspondence between properties measured on miniaturized libraries using high-throughput tools versus commercial-scale process sizes. This correspondence can place real limitations on industrial acceptance of CHTS as a tool for materiomics. Such comparisons are necessary in order to fully understand the costs and benefits of CHTS in materials research.

While integration of informatics and experimental approaches described above provides a promising start, the informatics tools required to further expand the applications of materiomics are still far from being complete. Biomaterials research perhaps started out with a slight advantage over general materials research, with regard to integrating modelling and experiments in high-throughput approaches, especially given the pre-existing history of combinatorial pharmaceutical development. However, the general materials field and the biomaterials field in particular are nearly at the same level of development regarding informatics, library preparation and property screening. Some authors have pointed to problems rooted in the current scholarly communication process and the way in which chemists and polymer scientists handle and publish data (59–63). For example, 'negative' data, such as experiments that did not work as expected, are often not published, whereas this information can be critical for data mining and future experimental design, to exclude structures and processes that do not generate desired results. Failures are less likely to be published than successes. Finally, commercial successes, and the methods used to discover them, may be delayed in publication owing to intellectual property and trade secret concerns.

While materiomics has been applied to many characterization tools, a few types of measurements still remain to be adapted completely to HT methods. One tool that is used widely in conventional measurements, for which we have discussed novel applications in CHTS above, is solid-state NMR, a powerful tool to investigate underlying molecular and physico-chemical properties. NMR imaging (i.e. MRI) provides a parallel means to assess the spatial distribution of NMR parameters in libraries, but remains largely undeveloped at this point. However, the hardware and software limitations associated with the MRI characterization of polymer films, which can exhibit a large variety of properties, will likely require development of experimental protocols on a case-by-case basis. It should be kept in mind that the tremendous progress in clinical MRI is essentially based on optimizing the technique for one single sample: the human body. Standard MRI for the characterization of polymer-based gradient-composition films appears to be suitable for evaluation of overall film quality provided sufficient molecular mobility is present. This limitation is much less of an issue for existing NMR spectroscopic methods. A modern NMR spectrometer with solid-state NMR capabilities (high power for rf pulses and fast analogue-to-digital converter) is capable of characterizing materials with underlying molecular mobilities ranging from those encountered in rigid crystalline phases to liquid melts.

10.4 Snapshot summary

- Complexity in material composition and processing leads to the need for materiomics in a wide range of fields including:
  - Energy: biofuels, fuel cells, solar cells, hydrogen storage and CO_2 capture;
  - Sustainability: conversion of biomass into chemicals, lightweight materials for transportation applications;
  - Advanced materials: microelectronics, organic electronics, polymers and adhesives.
- Combinatorial synthesis, high-throughput screening, and informatics form the basis for applying materiomics to the following:
  - Phase behaviour and mechanical properties;
  - Electronic, optical and semiconducting properties;
  - Transport and catalytic properties.
- Successful products developed using combinatorial materiomics include:
  - Propylene–ethylene speciality copolymers;
  - Phosphors for displays;
  - Polymers used in electronics.

Further reading

Beers KL, Douglas JF, Amis EJ *et al*. Combinatorial measurements of crystallization growth rate and morphology in thin films of isotactic polystyrene. Langmuir. 2003;**19**:3935–40.

Briceno, G, Chang H, Sun XD *et al*. A class of cobalt oxide magnetoresistance materials discovered with combinatorial synthesis. Science. 1995;**270**:273.

Brocchini, S, James K, Tangpasuthadol V *et al*. A combinatorial approach to polymer design. J Am Chem Soc. 1997;**119**:4553.

Dickinson TA, Walt DR, White J *et al*. Generating sensor diversity through combinatorial polymer synthesis. Analyt Chem. 1997;**69**:3413.

Hanak JJ. The 'multiple-sample concept' in materials research: synthesis, compositional analysis and testing of entire multicomponent systems. J Mater Sci. 1970;**5**:964.

Kennedy K, Stefansky T, Davy G *et al*. Rapid method for determining ternary-alloy phase diagrams. J Appl Phys. 1965;**36**:3808–10.

Klein J, Lehmann CW, Schmidt HW *et al*. Combinatorial material libraries on the microgram scale with an example of hydrothermal synthesis. Angew Chem Int Ed. 1998;**37**:3369.

Meredith JC, Karim A, Amis *et al*. High-throughput measurement of polymer blend phase behavior. Macromolecules. 2000;**33**:5760–2.

Meredith JC, Smith AP, Karim A *et al*. Combinatorial materials science: thin-film dewetting. Macromolecules. 2000;**33**(26):9747–56.

Xiang X-D, Sun X, Briceno, G *et al*. A combinatorial approach to materials discovery. Science. 1995;**268**:1738.

References

1. Smith AP, Douglas J, Meredith JC, Karim A, Amis EJ. Combinatorial study of surface pattern formation in thin block copolymer films. Phys Rev Lett. 2001;**87**:15503–6.

2. Wang J, Yoo Y, Gao C *et al*. Identification of a blue photoluminescent composite material from a combinatorial library. Science. 1998;**279**:1712–14.

3. Davis RD, Lyon RE, Takemori MT, Eidelman N. High throughput techniques for fire resistant materials development. In: Wilkie CA, Morgan AB, eds. *Fire Retardancy of Polymeric Materials*. CRC Press; 2010. p. 421–51.

4. Lewis GJ, Sachtler JWA, Lowa JJ *et al*. High throughput screening of the ternary $LiNH_2$–MgH_2–$LiBH_4$ phase diagram. J Alloys Compounds. 2007;**446**–447:355–9.

5. Guerin S, Hayden BE, Lee CE, Mormiche C, Russell AE. High throughput synthesis and screening of ternary metal alloys for electrocatalysis. J Phys Chem B. 2006;**110**:14355–62.

6. Todd ADW, Mar RE, Dahn JR. Combinatorial study of tin-transition metal alloys as negative electrodes for lithium-ion batteries. J Electrochem Soc. 2006;**153**:A1998–A2005.

7. Potyrailo RA, Chisholm BJ, Olson DR, Brennan MJ, Molaison CA. Development of combinatorial chemistry methods for coatings: high-throughput screening of abrasion resistance of coatings libraries. Analyt Chem. 2002;**74**:5105–11.

8. Baerns M, Holeňa M. *Combinatorial Development of Solid Catalytic Materials*. Imperial College Press; 2009.

9. Keep A, Ellis S, Ball S. Combinatorial and high-throughput discovery and optimization of catalysts and materials. Platinum Metals Rev. 2007;**51**:204–7.

10. Seyler M, Stoewe K, Maier WF. New hydrogen-producing photocatalysts – a combinatorial search. Appl Catal B Envir. 2007;**76**(1–2):146–57.

11. Potyrailo R, Rajan K, Stoewe K *et al.* Combinatorial and high-throughput screening of materials libraries: review of state of the art. ACS Combinat Sci. 2011;**13**(6):579–633.

12. Jandeleit B, Schaefer DJ, Powers TS, Turner HW, Weinberg WH. Combinatorial materials science and catalysis. Angew Chem Int Ed. 1999;**38**:2494–532.

13. Meredith JC. A current perspective on high-throughput polymer science. J Mater Sci. 2003;**38**(22):4427–37.

14. Cawse JN, ed. *Experimental Design for Combinatorial and High-throughput Materials Development.* Wiley-Interscience; 2003.

15. Potyrailo RA. Sensors in combinatorial polymer research. Macromol Rapid Commun. 2004;**25**(1):78–94.

16. Hagemeyer A, Strasser P, Volpe AF, eds. *High-throughput Screening in Chemical Catalysis.* Wiley-VCH; 2004.

17. Amis EJ. Reaching beyond discovery. Nat Mater. 2004;**3**(2):83–5.

18. Meier MAR, Hoogenboom R, Schubert US. Combinatorial methods, automated synthesis and high-throughput screening in polymer research: The evolution continues. Macromol Rapid Commun. 2004;**25**(1):21–33.

19. Meier MAR, Schubert US. Combinatorial polymer research and high-throughput experimentation: powerful tools for the discovery and evaluation of new materials. J Mater Chem. 2004;**14**(22):3289–99.

20. Zhang HQ, Hoogenboom R, Meier MAR, Schubert US. Combinatorial and high-throughput approaches in polymer science. Measure Sci Technol. 2005;**16**(1):203–11.

21. Meier MAR, Schubert US. Selected successful approaches in combinatorial materials research. Soft Matter. 2006;**2**(5):371–6.

22. Webster DC, Chisholm BJ, Stafslien SJ. Mini-review: Combinatorial approaches for the design of novel coating systems. Biofouling. 2007; **23**(3):179–92.

23. Adams N, Schubert US. Software solutions for combinatorial and high-throughput materials and polymer research. Macromol Rapid Commun. 2004;**25**(1):48–58.

24. Muster TH, Trinchi A, Markley TA *et al.* A review of high-throughput and combinatorial electrochemistry. Electrochim Acta. 2011;**56**(27):9679–99.

25. Kennedy K, Stefansky T, Davy G, Zackay VF, Parker ER. Rapid method for determining ternary-alloy phase diagrams. J Appl Phys. 1965;**36**:3808–10.

26. Hanak JJ. The 'multiple-sample concept' in materials research: synthesis, compositional analysis and testing of entire multicomponent systems. J Mater Sci. 1970;**5**:964.

27. Fodor SPA, Read JL, Pirrung MC *et al.* Light-directed, spatially addressable parallel chemical synthesis. Science. 1991;**251**:767–73.

28. Xiang X-D, Sun X, Briceno G *et al.* A combinatorial approach to materials discovery. Science. 1995;**268**:1738.

29. Mordkovich VZ, Jin Z, Yamadab Y *et al.* Fabrication and characterization of thin-film phosphor combinatorial libraries. Solid State Sci. 2002;**4**:779–82.

30. Großhans I, Karl H, Stritzker B. Advanced apparatus for combinatorial synthesis of buried II / VI nanocrystals by ion implantation. Mater Sci Eng. 2003;**B101**:212–15.

31. Takeuchi I, van Dover RB, Koinuma H. Combinatorial synthesis and evaluation of functional inorganic materials using thin-film techniques. MRS Bull. 2002;**27**:301–8.

32. Fasolka MJ, Stafford CM, Beers KL. Gradient and microfluidic library approaches to polymer interfaces. In: Meier MAR, Webster DC, eds. *Polymer Libraries 2010*. Springer; 2010. p. 63–105.

33. van den Berg AMJ, Smith PJ, Perelaer J *et al.* Inkjet printing of polyurethane colloidal suspensions. Soft Matter. 2007;**3**(2):238–43.

34. Kim KW, Jeon MK, Oh KS *et al.* Combinatorial approach for ferroelectric material libraries prepared by liquid source misted chemical deposition method. Proc Natl Acad Sci. 2007;**104** (4):1134–9.

35. Genzer J, Bhat RR. Surface-bound soft matter gradients. Langmuir. 2008;**24**:2294–317.

36. Meredith JC, Karim A, Amis EJ. High-throughput measurement of polymer blend phase behavior. Macromolecules. 2000;**33**:5760–2.

37. Meredith JC, Karim A, Amis EJ. Combinatorial and high-throughput methods in polymer science. MRS Bull. 2002;**27**(4):330–5.

38. Meredith JC, Smith AP, Karim A, Amis EJ. Combinatorial materials science: thin-film dewetting. Macromolecules. 2000; **33**(26):9747–56.

39. Gomez I, Basak P, Meredith JC. Polymer thickness and composition gradients. In: Genzer J, ed. *Soft Matter Gradient Surfaces: Methods and Applications*. Wiley; 2012.

40. Gallant ND, Lavery KA, Amis EJ, Becker ML. Universal gradient substrates for 'click' biofunctionalization. Adv Mater. 2007;**19**(7):965-+.

41. Gershon AL, Kota AK, Bruck HA. Characterization of quasi-static mechanical properties of polymer nanocomposites using a new combinatorial approach. J Compos Mater. 2009;**43** (22):2587–98.

42. Woodhouse M, Parkinson BA. Combinatorial approaches for the identification and optimization of oxide semiconductors for efficient solar photoelectrolysis. Chem Soc Rev. 2009;**38** (1):197–210.

43. Shinar J, Shinar R, Zhou Z. Combinatorial fabrication and screening of organic light-emitting device arrays. Appl Surface Sci. 2007;**254**:749–56.

44. Jayaraman S, Hillier AC. Electrochemical synthesis and reactivity screening of a ternary composition gradient for combinatorial discovery of fuel cell catalysts. Measure Sci Technol. 2005;**16**(1):5–13.

45. Zapata P, Mountz D, Meredith JC. High-throughput characterization of novel pvdf/acrylic polyelectrolyte semi-interpenetrated network proton exchange membranes. Macromolecules. 2012;**43**:7625–36.

46. Sormana JL, Chattopadhyay S, Meredith JC. High-throughput mechanical characterization of free-standing polymer films. Rev Sci Instrum. 2005;**76**(6):062214.

47. Sormana J-L, Meredith JC. High-throughput dynamic impact characterization of polymer films. Mater Res Innov. 2003;**7**(5):295–301.

48. Sormana J-L, Meredith JC. High-throughput screening of mechanical properties on temperature gradient polyurethaneurea libraries. Macromol Rapid Comm. 2003;**24**:118–22.

49. Hajduk D. Combinatorial polymer characterization. Abstr PAP ACS. 2001;**222**:338-POLY Part 2.

50. Tweedie CA, Anderson DG, Langer R, Van Vliet KJ. Combinatorial material mechanics: High-throughput polymer synthesis and nanomechanical screening. Adv Mater. 2005;**17** (21):2599.

51. Potyrailo RA, Wroczynski RJ, Lemmon JP, Flanagan WP, Siclovan OP. Fluorescence spectroscopy and multivariate spectral descriptor analysis for high-throughput multiparameter

optimization of polymerization conditions of combinatorial 96-microreactor arrays. J Comb Chem. 2003;**5**(1):8–17.

52. Gabriel C, Lilge D, Kristen MO. Automated raman spectroscopy as a tool for the high-throughput characterization of molecular structure and mechanical properties of polyethylenes. Macromol Rapid Comm. 2003;**24**:109–12.

53. Eidelman N, Simon CG. Characterization of combinatorial polymer blend composition gradients by FTIR microspectroscopy. J Res of Nat Inst Standards Technol. 2004;**109**(2):219–31.

54. Potyrailo RA, Wroczynski RJ, Lemmon JP, Flanagan WP, Siclovan OP. Fluorescence spectroscopy and multivariate spectral descriptor analysis for high-throughput multiparameter optimization of polymerization conditions of combinatorial 96-microreactor arrays. J Combinat Chem. 2003;**5**(1):8–17.

55. Isaacs ED, Marcus M, Aeppli G *et al*. Synchrotron X-ray microbeam diagnostics of combinatorial synthesis. Appl Phys Lett. 1998;**73**(13):1820.

56. Leisen J, Gomez I, Roper J, Meredith JC, Beckham H. Spatially resolved solid-state 1H NMR for evaluation of gradient-composition polymeric materials. ACS Combinat Sci. 2012;**14**:415–24.

57. Han S, Huang Y, Watanabe T *et al*. High-throughput screening of MOFs for CO_2 separation. ACS Combinat Sci. 2012;**14**:263–7.

58. Wollmann P, Leistner M, Stoeck U *et al*. High-throughput screening: speeding up porous materials discovery. Chem Commun. 2011;**47**:5151–3.

59. Broderick S, Suh C, Nowers J *et al*. Informatics for combinatorial materials science. JOM J Min Metals Mater Soc. 2008;**60**(3):56–9.

60. Kholodovych V, Smith J, Knight D *et al.* Accurate predictions of cellular response using QSPR: a feasibility test of rational design of polymeric biomaterials. Polymer. 2004;**45**:7367–79.

61. Adams N. Polymer Informatics. In: Meier MAR, Webster DC, eds. *Polymer Libraries 2010.* Springer; 2010. p. 107–49.

62. Moloshok TD, Klevecz RR, Grant JD *et al*. Application of Bayesian decomposition for analysing microarray data. Bioinformatics. 2002;**18**(4):566–75.

63. Zhang W, Fasolka M, Karim A, Amis E. An informatics infrastructure for combinatorial and high-throughput materials research built on open source code. Measure Sci Technol. 2005;**1**:261–9.

64. Suh C, Sieg SC, Heying MJ *et al.* Visualization of high-dimensional combinatorial catalysis data. J Combinat Chem. 2009;**11**:385–92.

65. Takeuchi I, Lippmaa M, Matsumoto Y. Combinatorial experimentation and materials informatics. MRS Bull. 2006;**31**:999–1003.

66. Kubinyi H. QSAR and 3D QSAR in drug design. Part 1: Methodology. Drug Discovery Today. 1997;**2**(11):457–67.

11 Beyond bed and bench

Tahir A. Mahmood and Jan de Boer

Materiomics is growing. In this book, we have discussed and reviewed how one side of the materiomics coin, the screening of libraries of biomaterials, is developing, but the future of screening will go hand in hand with the other side of the coin – modelling. Materiomics is growing and is, one could say, still in its infancy. Currently, most materiomics research is performed by academic research groups, although a small but increasing number of companies are active in this area. It will be helpful to understand how fast this field will grow, and how it might develop.

To sketch a potential scenario perhaps it is worth looking at recent developments in the field of molecular biology. In the mid-1990s, there were specialized machines in the lab to produce oligonucleotides, so-called primers, used in the polymerase chain reaction, a specialist job which gave the lab a scientific head start. These days, however, a primer can be ordered online and be at your bench in days, for the cost of a few dozen Euros. By the late 1990s, we saw cDNAs arrayed on nitrocellulose, which allowed simultaneous quantification of expression for dozens to hundreds of genes. No more tedious northern blotting one gene at a time! Driven by these early successes, many institutes started investing in arraying equipment and software development, as DNA microarray analysis became increasingly popular. But the real breakthrough in the use of this technology came when companies such as Affymetrix invested not only in standardization and professionalization of the technology, but importantly also in the work flow of a DNA microarray experiment. These companies provided a full package with not only arrays but also the incubators, washing, data analysis, detailed work protocols, image analysis software and a customer service platform. In both cases, the new technological possibilities of oligonucleotide synthesis and DNA microarray production paved the way for breakthroughs in science. We think that this will also be possible for materiomics.

A second important aspect in the genomics revolution was the reduction in cost. These days, one can perform a DNA microarray experiment for less than 100 Euros, which is about an eighth of the commercial prices a decade ago. Microarray experiments are now within the financial reach of many labs and are becoming more and more a standard molecular biology tool, rather than a scientific discipline of their own. Hence adaptation of this technology by the scientific community will be greatly aided when screening can become commoditized, or is something which can be outsourced.

Materiomics: High-Throughput Screening of Biomaterial Properties, ed. Jan de Boer and Clemens van Blitterswijk. Published by Cambridge University Press. © Cambridge University Press 2013.

The true power of materiomics will only be unleashed when it becomes available for every lab, not only the pioneering labs or those with a big enough budget to purchase expensive equipment. Standardization and commodity solutions are therefore necessary. It is to be expected that developments in micrototal analysis systems (μTAS) , using microfabricated analysis systems, can expedite the analysis of libraries of materials.

Another important factor in the development of this technology is adaptation of this approach by R&D departments of large companies. Here, we would like to draw a parallel with the field of therapeutic antibodies. With the breakthrough of the hybridoma technology, we are now able to produce antibodies against virtually any epitope, and similar to oligonucleotides, we can post-order virtually any antibody. Because each antibody binds to unique molecules in the cell, and binding sometimes results in inactivation of the molecule, the field started generating so-called therapeutic antibodies. For instance, the first therapeutic antibody to enter the market, muromonab-CD3 (Orthoclone CKT-3 by Ortho for reversal of kidney transplant rejection), was directed against the CD3 antigen of human T cells. Small companies such as Cambridge Antibody Technologies were spun out of academic groups and developed the technology to produce a truly novel way of treating disease. The strategy is effective and antibody therapeutics have become a big industry, but the large pharmaceutical companies only stepped in when the technology was validated and the risk had been substantially reduced by the smaller, more specialized companies. There are currently 26 approved monoclonal antibody therapeutics, and this is the second fastest growing segment of the global therapeutics market, after vaccines. A similar scenario may also play out in the field of materiomics. A few small companies have been started, based on the technology platforms conceived by their academic innovators; but will they be able to bring the screening technology to the market? The most important industry for biomaterials is the medical device industry – a large industry dominated by a few companies with broad portfolios, such as Johnson&Johnson and Medtronic. However, there are also hundreds if not thousands of smaller companies with narrower scopes that range from dental applications or calcium phosphate-based bone void fillers to contact lenses manufacturers. As described in Chapter 9, an implantable device is more than its chemistry or surface structure; it has a three-dimensional shape, produced using technology that may not be compatible with the manufacturing process of the original implant. It is therefore obvious that, in order to bring the benefits of materiomics forward to improve medical devices, close collaboration with established device manufacturers is essential. A critical aspect in this will be the first breakthrough, just as therapeutic antibodies were adapted by the pharmaceutical industry after the first noticeable successes. A smart material, selected from a library of thousands, and with an obvious advantage in efficacy or safety over current materials will serve to advertise the possibilities of this approach.

There are two very important factors for the slow rate of uptake of new technologies in the medical device industry. The first, as mentioned earlier, is the adaptation of the production process. While it is desirable to have a nano-topographical structure stimulating bone formation, how can this structure be imprinted on a titanium hip implant cast in a mould? Solutions can be found in microfabrication technology, but nonetheless require changing the production process. A second concern would be the regulatory pathway which has to be followed to introduce the new product into the market. Currently, the rules for medical device approval in developed countries are typically less onerous than for drugs. For small changes in the chemistry or architecture of an implant, such as a change in the manufacturing procedure of a bone void filler or sandblasting of a titanium hip implant, a 510(k) FDA registration would normally be deemed sufficient in the United States. It has historically been considered wise to attempt product development strategies that result in compliance with the 510(k) regulations. For example, one could think of blending of polymers and other biomaterials that have market approval, or finding surface structures with beneficial properties. Nevertheless, it is likely that the bioactive properties of the new generation of biomaterials will not go unnoticed by the regulatory bodies. Indeed, the development of advanced materials that have intrinsic capability to modulate their local biological environment is causing a paradigm shift in the medical device regulatory world. Recent guidance from senior leaders at the US Food and Drug Administration (FDA) has implied that primary responsibility for the review of medical devices such as implants used in orthopaedic or spinal surgery may soon be moved from the Center for Devices and Radiological Health (CDRH – traditionally the default division of the FDA for medical device review) to the Center for Biologics Evaluation and Research (CBER). CBER is the FDA division that oversees review of biological therapeutics such as growth factors, monoclonal antibodies and other non-immunoglobulin-based proteins. The fact that this division is being primed for authority over implantable devices that do not incorporate any sort of drug is indicative of the growing appreciation of the bioactive nature of many new implantable materials, which can interrogate the biological milieu like therapeutic agents, and therefore are deemed to require review that is comparable with their counterpart therapeutic products. As described in the introduction of this book, during the early days of the field, biomaterials were selected based on their inert nature. With the new focus of the field on bioactive materials, it is also plausible that anything that is bioactive is more likely to have side effects when implanted into the body than bioinert materials. An example of this is the field of nanotoxicology, a research field dedicated to potentially hazardous effects of particles with submicrometre sizes. These could arise from the use of nanoscale drug carriers, ferromagnetic beads for tumour heat ablation, or debris from wear on implants.

The examples mentioned above illustrate that the biomaterials field is evolving in parallel with other elements within society related to this space, such as policy, ethics, investment and economic growth. This is obvious in regulatory affairs, and means that both academic and industrial players will have to change the way they work. It also means that we need to pay attention to the way we train new scientists, as well as ourselves. This is not new. When the biologist James Watson dedicated himself to discovering the structure of DNA, he had to master at least the basics of X-ray crystallography to be able to discuss his ideas with his physics colleagues. Of course, this is not unique to materiomics. If we look at the use of ICT in biomedical research these days, we constantly need to educate ourselves about the latest developments. In fact, bioinformaticists are now as common in biomedical laboratories as in clinical centres, and the integration and analyses of genomic, metabolomic, proteomic and imaging data using cloud computing clusters are leading to large-scale research across organizations and even across countries. An example of this is the recent

Cancer Bioinformatics Grid (CaBIG) established by the National Cancer Institute of the US National Institutes of Health. Furthermore, the growth of wireless and mobile health technologies is revolutionizing translational research and the practice of medicine, in both developed and developing regions of the world. Young companies such as Proteus Digital Health and Airstrip are pioneering the application of low-radiofrequency emitting nano-fabricated 'chips' embedded in drug pills, and the transmission of medical readouts and waveforms, respectively. These technologies have added the complexity of physics, electrical engineering and nanofabrication to the already complex underlying pharmacological and physiological systems.

In addition to the use of 'smart' materials in the areas of tissue engineering and implantable devices, there are significant other opportunities in the areas of diagnostics and research tools. Companies such as Life Technologies, Illumina and Prognosys Biosciences are aggressively pushing the technology envelope in their respective fields, and would find it valuable to be able to use materials science and chemistry to elicit biological responses akin to those of bioactive agents that are more expensive and require cumbersome processes to incorporate and maintain. The application of materiomics to these areas would be likely to resemble the use of antibodies and proteins at the advent of the era of modern biochemical assays, and would represent a turning point in the commoditization of materiomics across multiple segments of the biomedical industry.

Going back to the field of biomaterials research, we see that tissue engineering has brought biologists and material scientists together. Traditionally, biology has been a fundamental science, and most biologists (but certainly not all) do not have a natural tendency to think as engineers – or as appliers of gleaned knowledge. In contrast, material sciences are often application-driven but the scientists involved may have limited background in biology. To tackle this problem, many universities have initiated academic programmes on biomedical technology and biomedical engineering, where the students are offered a mixed curriculum of topics such as chemistry, imaging, biology and nanotechnology. Similarly, the increasing use of technology in hospitals has called for a new generation of doctors whose basic training is in medicine, but who will specialise in 'technical medicine', in order to facilitate the interface between technology and medicine. With the development of the biomaterials field towards high-throughput screening and modelling, and the subjects described in this book, ranging from microfabrication to computational modelling and clinical implementation, we should think about how to train ourselves and our students. Do we train generalist technologists, or scientists with in-depth knowledge in classical disciplines who are able to collaborate, just like Watson did? It takes a polymer chemist to produce a new block copolymer system, but it takes a mechanical engineer, a polymer chemist and a biologist to produce a library of block copolymers and screen it with cells. In fact, it takes a biologist with keen interest in biomaterials research and a polymer chemist with an interest in the biomedical possibilities of their polymers to take this research to the next level.

Through the content of this book, we hope to inspire people to be experts in their own disciplines, but also to be brave enough to look into other scientific disciplines and to be able to communicate effectively with their practitioners.

Index